T0135613

The Effect of a Singular Perturbation
to a 1–d Non–Convex Variational Problem

Dissertation
von Markus Lilli

eingereicht
beim Institut für Mathematik
der Mathematisch-Naturwissenschaftlichen Fakultät
der Universität Augsburg
im November 2004

Augsburger Schriften zur Mathematik, Physik und Informatik
Band 10

herausgegeben von:
Professor Dr. F. Pukelsheim
Professor Dr. W. Reif
Professor Dr. D. Vollhardt

Bibliografische Information Der Deutschen Bibliothek

Die Deutsche Bibliothek verzeichnet diese Publikation in der Deutschen
Nationalbibliografie; detaillierte bibliografische Daten sind im Internet über
http://dnb.ddb.de abrufbar.

ISBN 3-8325-0928-3
ISSN 1611-4256

Logos Verlag Berlin
Comeniushof, Gubener Str. 47,
10243 Berlin
Tel.: +49 030 42 85 10 90
Fax: +49 030 42 85 10 92
INTERNET: http://www.logos-verlag.de

Vorwort

Nichtkonvexe Variationsprobleme stellen nach wie vor eine mathematische Herausforderung dar. Die fehlende Konvexität lässt die direkten Methoden der Variationsrechnung scheitern, was insbesondere bedeutet, dass auch bei Beschränktheit nach unten die Existenz eines Minimierers offen ist. Klassische Beispiele, von denen noch zu sprechen sein wird, lassen keine Minimierer zu, andere wiederum besitzen welche, so dass eine abgeschlossene Theorie nicht absehbar ist. Die Schwierigkeiten sind naturgemäß für höher–dimensionale Modelle größer, sie treten aber auch schon für den 1–dimensionalen Fall auf. Alle diese nichtkonvexen Funktionale sind als Phasenübergangsmodelle von Interesse. Auf diese Anwendungen wird in der vorliegenden Arbeit allerdings nicht eingegangen. Sie werden nur erwähnt, um die mathematische Beschäftigung mit ihnen zu rechtfertigen.

Herr Lilli untersucht ein Energiefunktional über dem Intervall $[0,1]$, welches aus der Summe eines Zweitopfpotentials $W = W(x, u')$ und eines allgemeinen Potentials $B = B(x, u)$, wobei u eine hinreichend glatte Funktion bezeichne. Das Vorzeichen des Potentials B ist indessen entscheidend für die Nichtexistenz von Minimierern: Nehmen wir zum Beispiel positive Potentiale mit minimalen Werten Null, welche $W = W(u')$ bei ± 1 und $B = B(u)$ bei 0 annimmt, können keine Minimierer existieren, da die minimale Energie $u = 0$ und $u' = \pm 1$ bedingt, was sich natürlich ausschließt. Dennoch sind diese Modelle für Mikrostrukturen von Interesse: Minimalfolgen können zwar nicht gegen eine minimierende Funktion, wohl aber gegen ein sogenanntes Youngsches Maß konvergieren. Dieses ist die Wahrscheinlichkeitsverteilung der Werte von u' im Intervall $[0,1]$ und es gibt auf diese Weise Aufschluß über die Mikrostruktur des Modells. Bei Scheitern der direkten Methoden der Variationsrechnung ist der Zugang über die Euler–Lagrange Gleichung gerechtfertigt. Dummerweise ist diese für nichtkonvexe Variationsprobleme nicht elliptisch sondern wechselt den Typ, was bekanntlich zu großen mathematischen Problemen führt. Wie im Titel der Dissertation angekündigt, bedient sich Herr Lilli einer elliptischen Regularisierung, welche, lässt man sie wieder verschwinden, auch als singuläre Störung bezeichnet werden kann. Physikalisch lässt sie sich als vernachlässigte Energie der sprunghaften Übergänge von $u'(x) = p_1(x)$ zu $u'(x) = p_2(x)$ deuten, wobei $p_1(x), p_2(x)$ die Maxwell–Punkte von $W(x, \cdot)$ bezeichne, und sie ist von der Form $\varepsilon(u'')^2$ mit einem kleinen $\varepsilon > 0$. Mit diesem Zusatzkern wird die Euler–Lagrange–Gleichung ein elliptisches Randwertproblem 4.Ordnung.

Im ersten Teil seiner Arbeit untersucht Herr Lilli dieses Randwertproblem in Abhängigkeit von $\varepsilon > 0$. Sein schönes Resultat ist das folgende: Beim Grenzübergang $\varepsilon \searrow 0$ erzeugen die Lösungen $u = u_\varepsilon$ ein Youngsches Maß, welches eine verallgemeinerte Lösung der Euler–Lagrange–Gleichung 2.Ordnung für $\varepsilon = 0$ ist. Ist dieses ein Dirac–Maß, so kann man es als Funktion interpretieren, und in Abhängigkeit von W und B ist Herr Lilli in der Lage, den Träger dieses Youngschen Maßes anzugeben. Meines Wissens ist dies die erste allgemeine Lösungstheorie für Euler–Lagrange–Gleichungen im Raum der Wahrscheinlichkeitsmaße.

Im zweiten Teil greift Herr Lilli die Idee von Kielhöfer/Healey auf, Lösungen der Euler–Lagrange–Gleichung mit Hilfe einer Kontinuitätsmethode zu studieren. In diesem Fall hängt das Potential B von einem reellen Parameter λ ab, $B = B(\lambda, x, u)$, wobei die Kenntnisse über Lösungen für $\lambda = 0$ auf solche für positive λ übertragen werden. Von Interesse sind neben der Existenz auch qualitative Eigenschaften der Lösungen. Auch hier wird zuerst die singuläre Störung für $\varepsilon > 0$ und anschließend der Grenzübergang $\varepsilon \searrow 0$ untersucht. Während Kielhöfer/ Healey dieses Programm nur für spezielle B mit der Vorzeichenbedingung $B_u(u) \geq 0$ durchziehen konnten, ist dies Herr Lilli für allgemeinere Potentiale B mit einem Vorzeichenwechsel von B_u gelungen. Die qualitativen Eigenschaften der Lösungen u_ε werden mit Hilfe subtiler Maximumsprinzipien über alle λ "homotopiert", was eine fundamentale Kenntnis der qualitativen Theorie elliptischer Randwertprobleme 4.Ordnung voraussetzt. Man beachte, dass Maximumsprinzipien im allgemeinen nur für elliptische Probleme 2.Ordnung gelten. Auch hier wird der Grenzübergang $\varepsilon \searrow 0$ vollzogen: Im Gegensatz zum ersten Teil der Arbeit konnte Herr Lilli beweisen, dass der singuläre Grenzwert stets eine Funktion ist. Diese hat einen Sprung in der Ableitung, der mit Hilfe der Weierstrass–Erdmannschen Eckenbedingungen genau quantifiziert werden kann. Dieser Grenzwert löst die Euler–Lagrange–Gleichung 2.Ordnung für $\varepsilon = 0$, und für spezielle Potentiale B ist diese Lösung gleichzeitig der globale Minimierer.

Im dritten Teil der Dissertation untersucht Herr Lilli die Stabilität von Lösungen, genauer gesagt beantwortet er folgende Fragen: Ist der singuläre Grenzwert stabiler Lösungen der Euler–Lagrange–Gleichung 4.Ordnung für $\varepsilon = 0$ eine Funktion, welche eine stabile Lösung der Euler–Lagrange–Gleichung für $\varepsilon = 0$ ist? Welche qualitativen Eigenschaften haben diese stabilen Grenzwerte? Naturgemäß können diese Fragen nicht für allgemeine Potentiale B beantwortet werden, wohl aber für lineares $B(x, u) = b(x)u$ mit der Funktion b, welche das Vorzeichen wechseln kann. Unter weiteren Voraussetzungen zeigt Herr Lilli auch, dass der singuläre Grenzwert nicht nur stabil sondern sogar der globale Minimierer ist. Durch die Weierstrass–Erdmannschen Eckenbedingungen sind die Sprünge der Ableitung quantifiziert. Für globale Minimierer der singulären Störung kann ausserdem die Konvergenzrate genauer bestimmt werden. Zum Schluß gibt Herr Lilli noch eine Bedingung an, unter der globale Minimierer eindeutig sind. Der Beweis bedient sich des Gradientenflusses und der qualitativen Theorie unendlich–dimensionaler Systeme. Erfunden hat Herr Lilli diese Vorgehensweise nicht, er wendet sie aber geschickt für seine Zwecke an.

Augsburg, den 05.September 2005

Prof. Dr. Hansjörg Kielhöfer
Institut für Mathematik
Mathematisch–Naturwissenschaftliche Fakultät
Universität Augsburg

Abstract:

We consider a one–dimensional nonconvex variational problem and we discuss the corresponding Euler–Lagrange equation. Since this equation is not elliptic we consider a singular perturbed variational problem whose Euler–Lagrange equation is a fourth order elliptic equation having a solution by the direct methods of the calculus of variations. This work is separated into three parts: In the first and the second one we prove existence of solutions of the Euler–Lagrange equation of the nonconvex variational problem. First we prove existence of Young–measure solutions by assuming very general conditions on the potential. Moreover we give some results concerning the support of the Young–measure solution. By imposing more restrictive conditions on the potential in the second chapter, existence of classical solutions is proved by applying a continuation method. Also in this case we can characterize the solution by showing that it satisfies the first and second Weierstrass–Erdmann corner condition. In both cases we consider solutions of the Euler–Lagrange equation of the singular perturbed problem and pass to the limit.The third chapter is devoted to a special potential where the existence of a global minimizer of the non–convex problem is obvious. In this case we are mainly interested in the qualitative behavior of global and local minimizers of the singular perturbed problem.

Zusammenfassung:

Wir betrachten ein 1–d nichtkonvexes Variationsproblem und analysieren die zugehörige Euler–Lagrange Gleichung. Da diese Gleichung nichtelliptisch ist, betrachten wir das singulär gestörte Variationsproblem dessen Euler–Lagrange Gleichung eine vierte Ordnungs Gleichung ist und die aufgrund der direkten Methoden der Variationsrechnung eine Lösung besitzt. Diese Arbeit ist in drei Teile unterteilt: Im ersten wie im zweiten Kapitel beweisen wir die Existenz von Lösungen der Euler–Lagrange Gleichung des nichtkonvexen Variationsproblems. Zuerst beweisen wir Existenz von Lösungen im Raum der Youngschen Maße unter sehr allgemeinen Bedingungen an das Potential. Darüberhinaus charakterisieren wir den Träger des Youngschen Maßes. Unter schärferen Bedingungen an das Potential beweisen wir im zweiten Kapitel Existenz von klassischen Lösungen via topologischer Methoden. Zudem zeigen wir, dass die Lösungen die erste wie die zweite Weierstrass–Erdmannsche Eckenbedingung erfüllen. In beiden Fällen betrachten wir die Gleichung des singulär gestörten Problems und gehen zum Limes über. Im dritten Kapitel wird ein derartiges Potential untersucht, so dass die Existenz eines globalen Minimierers des nichtkonvexen Problems trivial ist. Hier sind wir vor allem am qualitativen Verhalten von globalen wie lokalen Minimierern des singulär gestörten Problems interessiert.

MSC: 35B05, 35B38, 35B25, 74B20

Contents

1 Introduction   **7**
   1.1 The model 7
   1.2 Outline 9

2 The Young measure approach   **11**
   2.1 Introduction 11
   2.2 Young measure preliminaries 12
   2.3 Model and basic definition 14
   2.4 A–priori bounds 16
   2.5 Results 17
   2.6 Concluding remarks 21

3 The continuation method approach   **23**
   3.1 Introduction 23
   3.2 Global analysis 25
   3.3 Properties of the continuum 29
   3.4 Some a–priori bounds 33
   3.5 Geometric structure of $\mathcal{C}^+$ 35
   3.6 Singular limit analysis 45
   3.7 The Limiting set is a continuum 53
   3.8 Minimizing properties of solutions
        for dead loading 54

4 Stability analysis   **55**
   4.1 Introduction 55
       4.1.1 The model 57
   4.2 Existence of a minimizer for
        the unrelaxed problem 58
   4.3 Compactness of a sequence of stable critical points 62
   4.4 Shape of stable critical points 66
   4.5 Asymptotic Behavior 79
   4.6 Necessary condition for uniqueness of the global Minimizer 84
       4.6.1 The linearized Problem 89
       4.6.2 Necessary condition for uniqueness 95

Bibliography   **101**

Index   **104**

Chapter 1

Introduction

In this thesis we consider a one–dimensional elastic bar placed in a soft loading device. Our approach is based on a singular perturbation of the nonconvex variational problem and considering the corresponding Euler–Lagrange equation. There is a vast literature concerning this and related topics, see for example [8], [9], [18],[25], [26], [27] and [33]. First we introduce the physical model of static elastic bars and therefore we follow the lines of [23].

1.1   The model

We consider an elastic bar placed in a soft loading device with a potential B delivering a body–force b. A soft loading device means, that the bar is fixed at one end of the bar (without loss of generality we assign the left end) and the right end of the bar is free. Let 1 be the length of the undeformed bar and let $[0, 1]$ be the bar interval. The soft loading device causes the boundary value $u(0) = 0$.

We call $u = u(x)$, $x \in [0, 1]$, the placement of the bar, which assigns the position of the bar at x. Furthermore let $u'(x)$ the deformation of the bar at x and $u'(x) - 1$ the strain.

We introduce a strain energy function $W(x, p)$ and the stress is defined by

$$\sigma(x, u') = \frac{\partial W}{\partial p}(x, u').$$

From a physical point of view, either cross–sections with identical elasticities have different sizes or shapes, or cross–sections of identical size and shape have different elasticities. Also combinations of them are possible. This is reflected by an explicit dependence of the strain energy function W upon the spatial variable x in the bar interval $[0, 1]$. The function $W(x, \cdot)$ is a double–well potential and in general it is just assumed to be two times continuously differentiable with respect to the second variable and measurable with respect to the first. Hence we obtain existence of points $c_1(x)$ and $c_2(x)$ such that

$$W_{pp}(x, p) \begin{cases} < 0, & \text{if } p \in (c_1(x), c_2(x)) \\ > 0, & \text{if } p \in (-\infty, c_1(x)) \cup (c_2(x), \infty). \end{cases}$$

A typical W for fixed x is shown in the picture:

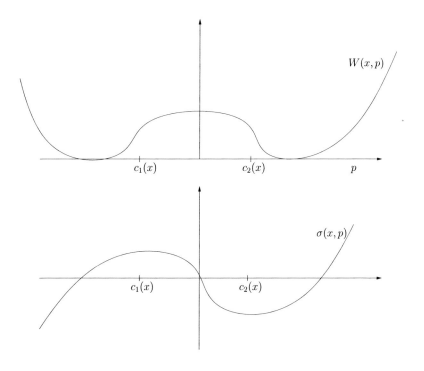

Furthermore let $B(u, x)$ a potential delivering in general a live body force

$$-\frac{\partial B}{\partial u}(u, x) := b(u, x).$$

Moreover the soft loading device produces a load at $x = 1$ given by a certain constant t_r and commonly such a load is called a dead load.

The total energy of the bar is defined by

$$\tilde{J}(u) := \int_0^1 W(x, u') + B(u, x)\, dx - \int_1^{u(1)} t_r\, ds,$$

$$u(0) = 0.$$

Integration yields

$$\tilde{J}(u) := \int_0^1 W(x, u') + B(u, x)\, dx - u(1)\, t_r + t_r.$$

Due to the fact that the last term is just a constant, it is from the mathematical point of view enough to consider

$$J(u) := \int_0^1 W(x, u') + B(u, x)\, dx - u(1)\, t_r, \tag{1.1}$$

$$u(0) = 0.$$

Our goal is to study the fully nonlinear Euler–Lagrange equation corresponding to (1.1) by three different approaches and in each of them we have to specify our conditions on W and B. In every approach we use an additional interfacial energy to obtain the following variational problem:

$$J(u) := \int_0^1 \frac{\varepsilon}{2}u''^2 + W(x, u') - B(u, x)\, dx - u(1)\, t_r, \tag{1.2}$$

$$u(0) = 0.$$

The corresponding semilinear EL equation can be denoted as a fourth order equation

$$-\varepsilon u^{(4)} + \frac{d}{dx}\left[W'(x, u')\right] + b(u, x) = 0 \tag{1.3}$$

with boundary conditions

$$u(0) = u''(0) = u''(1) = 0, \qquad \varepsilon u'''(1) = W'(u'(1)) - t_r.$$

1.2 Outline

The variational problem (1.1), the singular perturbed functional (1.2) and the corresponding Euler–Lagrange equation (1.3) are discussed by using several tools:

In our first approach we prove existence of Young–measure solutions of the fully–nonlinear Euler–Lagrange equation of the variational problem (1.1) under very mild assumptions on W and B. These generalized solutions are generated by classical solutions of (1.3). Moreover we calculate the Young measure in certain regions D (namely those in which the stress is injective). In particular it turns out that this Young measure consists of Dirac measures for almost every $x \in D$. Furthermore it seems remarkable to us, that the 1–dimensional case allows us to define a Young measure solution in a pointwise sense and not only in the distributional one. In particular, this solution satisfies the first Weierstrass–Erdmann corner condition in a generalized sense.

The second approach is done by global continuation. In this case we have to restrict ourselves to a homogenous "one–well" potential W and some (in general live) body force changing sign once in a certain assigned manner. Moreover we set $t_r = 0$. This approach yields existence of (classical) solutions $u \in W^{1,\infty}(0,1) \cap \{v \mid v' \in BV(0,1)\}$ of the fully nonlinear Euler–Lagrange equation corresponding to (1.1) by proving convergence in $W^{1,1}(0,1)$ of a sequence of solutions of (1.3) as $\varepsilon \searrow 0$. We remark, that we cannot prove that these solutions are global minimizers of (1.1) (besides in the dead load case), but nevertheless we conjecture that because every solution we obtain by our method satisfies the first and second Weierstrass–Erdmann corner conditions.

In the last approach we do not emphasize on existence of solutions of the Euler–Lagrange equation of (1.1) (existence of a global minimizer u_* is in this case trivial by our assumptions on the body–force), it rather deals with stable solutions of (1.3). We prove, that under appropriate assumptions on b every sequence of stable solutions converge to u_* in $W^{1,1}(0,1)$. Moreover we detect some features of them for ε sufficiently small, especially the width of a transition layer depending on ε. Uniform convergence to u_*' is proved in regions where no jump of this function occurs. Furthermore we give a necessary condition for uniqueness of the global minimizer of (1.2) in terms of the position of its transition layer. In this approach we consider a homogenous W and a dead load b (hence b is independent of u) which is allowed to change sign finitely many times. However, we can allow traction at the right end of the bar.

Acknowledgements: I am indepted to my supervisors Prof. Hansjörg Kielhöfer and Prof. Timothy J. Healey for numerous valuable suggestions and hints as well as for many helpful discussions. Especially I want to thank Prof. Healey for his invitation to the Department of Theoretical and Applied Mechanics at Cornell University, where I had the opportunity to present my results. Furthermore I want to express my gratitude to Stefan Krömer and Niko Tzoukmanis for profound discussions. Finally, I want to thank the Graduiertenkolleg "Nichtlineare Probleme in Analysis, Geometrie und Physik" at the University of Augsburg for financial support.

Chapter 2

The Young measure approach

2.1 Introduction

In recent years Young measures became one of the most important tools for describing oscillating structures associated with a sequence of functions. Especially it was used in non–convex problems in calculus of variations in which usually no minimizer exist in a classical function space. Typically an arbitrary minimizing sequence causes finer and finer oscillations and this prevents strong convergence in some Sobolev–space. These oscillations can be described by using Young measures. Thus the key idea was to consider the variational problem over the space of Young measures which is in fact an extension of some suitable Sobolev–space $\mathcal{A}$. Actually it turns out, that under appropriate growth conditions on the potential the variational problem has a global minimizer over this space and moreover the problem possesses also a global minimizer in $\mathcal{A}$, if and only if the Young measure is a Dirac measure. For a detailed treatment of this topic we refer to [35].

In the subject of partial differential equations the Young measure was as far as we know introduced by Tartar as a suitable expression of a solution. The structure of measure–valued solutions encodes the type of oscillations and also the character of convergence that is realized by the generating sequence.

In [40] he gave a proof for existence of a classical solution of the scalar conservation law

$$\frac{\partial u}{\partial t} + \frac{\partial}{\partial x} F(u) = 0 \qquad (2.1)$$

in one space dimension by showing that the viscosity solutions of (2.1) generate a Young measure which turns out to be a Dirac measure for almost every x. This last step is the core of the paper and it is deduced by a method called compensated compactness. Tartar's proof requires only bounds on the amplitude of u_ε but no a–priori estimate is needed for the derivatives.

A detailed survey concerning Young measures and measure valued solutions for PDE's is given in [31].

Other papers, where also Young measure solutions are obtained for quasilinear elliptic equations are [13], [22] and [30]. By a similar approach like in this work they were able to prove that the Young measure consists of Dirac measures. In recent years the dynamic behavior of microstructures became more and more important. Papers regarding that topic are for example [28], [38] and [41].

As far as we know we present in this paper the first treatment of a static Euler–Lagrange equation emphasizing on Young–measure solutions. We restrict ourselves to the one dimensional case which allows us to strengthen the notion of a Young–measure solution in a very effective way. Usually the Young–measure solution of a given PDE is defined in the sense of distributions and we state for example the definition given in [12] for the scalar hyperbolic equation

$$\partial_t u + \partial_x f(u) = 0 : \qquad (2.2)$$

A measurable map

$$\nu : y \mapsto \nu_y \in \text{Prob}(\mathbb{R}^n)$$

from the domain $\mathbb{R}^2$ to the space of probability measures over $\mathbb{R}^n$ is a Young–measure solution of (2.2) if

$$\partial_t \langle \nu_y, Id \rangle + \partial_x \langle \nu_y, f \rangle = 0$$

in the sense of distributions on $\mathbb{R}^2$, i.e.

$$\int \int \left[\langle \nu_y, Id \rangle \, \phi_t + \langle \nu_y, f \rangle \, \phi_x \right] dx \, dt = 0 \ \text{ for every } \phi \in C_0^1(\mathbb{R}^2).$$

The definition of $\langle \, , \rangle$ is given by (2.3).
In our case the goal is to obtain a Young–measure solution satisfying the Euler–Lagrange equation of a given 1–dimensional variational problem pointwise.

First we introduce the notion of Young measures, the model under consideration and the basic definition for a Young measure solution.

2.2   Young measure preliminaries

We just give a brief introduction into the theory of Young measures. For a detailed discussion we refer to [7], [33] and [35].
First we have to introduce some function–spaces:
Let Ω be an open and bounded subset in $\mathbb{R}^n$, let $C^0(\mathbb{R}^d)$ be the space of continuous functions defined over $\mathbb{R}^d$ and let $\mathcal{M}(\mathbb{R}^d)$ denote the space of $\mathbb{R}$–valued Radon–measures with support on $\mathbb{R}^d$, which can be viewed as the dual of

$$C_0(\mathbb{R}^d) := \left\{ \varphi \in C^0(\mathbb{R}^d) \mid \lim_{|x| \to \infty} \varphi(x) = 0 \right\}$$

via the dual pairing

$$\langle \nu, f \rangle = \int_{\mathbb{R}^d} f \, d\nu. \qquad (2.3)$$

The mapping $\nu : \Omega \to \mathcal{M}(\mathbb{R}^d)$ is called weak* measurable, if the function $x \mapsto \langle \nu(x), f \rangle$ is measurable for every $f \in C_0(\mathbb{R}^d)$. Throughout the whole chapter we will denote $\nu(x) := \nu_x$.
The space $L_w^\infty(\Omega, \mathcal{M}(\mathbb{R}^d))$ consists of all weakly* measurable mappings $\nu : \Omega \to \mathcal{M}(\mathbb{R}^d)$

which are essentially bounded and furthermore we endow the space with the weak*
topology induced by duality with $L^1(\Omega, C_0(\mathbb{R}^d))$. (See [14] for details). That means

$$\nu^n \overset{*}{\rightharpoonup} \nu \quad \text{in } L_w^\infty(\Omega, \mathcal{M}(\mathbb{R}^d)), \tag{2.4}$$

if and only if

$$\int_\Omega \langle \nu_x^n, f \rangle \, g(x) dx \to \int_\Omega \langle \nu_x, f \rangle \, g(x) dx \quad \text{for every } f \in C_0(\mathbb{R}^d) \text{ and } g \in L^1(\Omega).$$

The fundamental theorem on Young measures was proven by Ball (see [7]) and we state
here a version given in [33]:

Theorem 1. *Let $E \subseteq \mathbb{R}^n$ be a measurable set of finite measure and let $z_j : E \to \mathbb{R}^d$
be a sequence of measurable functions. Then there exists a subsequence $(z_{j_k})_{k \in \mathbb{N}}$ and a
weak* measurable map $\nu : E \to \mathcal{M}(\mathbb{R}^d)$ such that the following holds:*

(i)

$$\nu_x \geq 0, \quad ||\nu_x||_{\mathcal{M}(\mathbb{R}^d)} := \int_{\mathbb{R}^d} d\nu_x \leq 1, \quad \text{for almost every } x \in E.$$

(ii) *For every $f \in C_0(\mathbb{R}^d)$ we have*

$$f(z_{j_k}) \overset{*}{\rightharpoonup} \bar{f} \quad \text{in } L^\infty(E),$$

where

$$\bar{f}(x) = \langle \nu_x, f \rangle = \int_{\mathbb{R}^d} f \, d\nu_x.$$

(iii) *Furthermore one has*

$$||\nu_x||_{\mathcal{M}(\mathbb{R}^d)} = 1 \quad \text{for a.e. } x \in E \tag{2.5}$$

if and only if the sequence is bounded in any L^p for every $p > 0$.

Remark 2. (i) The map ν is called the Young measure generated by the sequence
$(z_{j_k})_{k \in \mathbb{N}}$, if the sequence is uniformly bounded in some L^p, $p > 0$. In particular
ν_x is a probability measure for almost every $x \in \Omega$ due to (2.5).

(ii) To prove (ii) in Theorem 1 one has to show

$$\delta_{z_{j_k}} \overset{*}{\rightharpoonup} \nu \quad \text{in } L_w^\infty(\Omega, \mathcal{M}(\mathbb{R}^d))$$

in the sense described in (2.4), which is an immediate consequence of Banach–
Alaoglu's theorem applied to the separable Banach space $L_w^\infty(\Omega, \mathcal{M}(\mathbb{R}^d))$.

An easy consequence of Theorem 1 is the following:

Corollary 3. *Let $(z_j)_{j \in \mathbb{N}}$ be a sequence uniformly bounded in $C^0(E, \mathbb{R}^d)$, let ν be
the Young measure generated by a subsequence of $(z_j)_{j \in \mathbb{N}}$ (not relabelled) and let $f \in
C^0(\mathbb{R}^d)$. Then we get*

$$f(z_j) \overset{*}{\rightharpoonup} \langle \nu_x, f \rangle \quad \text{in } L^\infty(E).$$

In particular, let $z \in L^\infty(E, \mathbb{R}^d)$ such that

$$z_j \overset{*}{\rightharpoonup} z \quad \text{in } L^\infty(E, \mathbb{R}^d)$$

and let $f = Id$, then we obtain

$$\langle \nu_x, Id \rangle = z(x) \quad \text{for a.e. } x \in E.$$

Furthermore we have the following useful fact. For a proof we also refer to [33]:

Corollary 4. *Suppose that a sequence* $(z_j)_{j \in \mathbb{N}}$ *of measurable functions from E to $\mathbb{R}^d$ generates the Young measure* $\nu : E \to \mathcal{M}(\mathbb{R}^d)$. *Then*

$$z_j \to z \text{ in measure if and only if } \nu_x = \delta_{z(x)} \text{ a.e.,}$$

where f_j converges to f in measure means, that for every $\varepsilon > 0$ we have

$$\lim_{j \to \infty} |\{|f_j - f| \geq \varepsilon\}| = 0.$$

2.3   Model and basic definition

In this chapter we consider the one dimensional elastic bar placed in a soft loading device. We assume the bar to be inhomogenous which is from the physical point of view very reasonable for two reasons denoted in the introduction. Thus the strain energy function W depends explicitly upon the spatial variable x in the bar interval $[0,1]$. Furthermore W depends on u', the deformation of the bar, where u is its placement and W is assumed to be two times continuously differentiable with respect to the second variable and C^1 with respect to the first. The function $W(x, \cdot)$ is a double–well potential. Thus we assume existence of points $c_1(x)$ and $c_2(x)$ such that

$$W_{pp}(x, p) \begin{cases} < 0, & \text{if } p \in (c_1(x), c_2(x)) \\ > 0, & \text{if } p \in (-\infty, c_1(x)) \cup (c_2(x), \infty). \end{cases} \tag{2.6}$$

Furthermore let $B(u, x)$ a potential delivering in general a live body force

$$-\frac{\partial B}{\partial u}(u, x) := b(u, x),$$

where

$$b \in C(\mathbb{R} \times [0, 1]).$$

The left end of the bar is fixed which assigns the boundary condition $u(0) = 0$ and at the free end of the bar a traction is produced given by the constant t_r. The total energy of the bar is defined by

$$J(u) := \int_0^1 [W(x, u') + B(u, x)] \, dx - t_r u(1), \tag{2.7}$$
$$u(0) = 0.$$

The corresponding Euler–Lagrange equation is

$$\frac{d}{dx}[W_p(x, u')] + b(u, x) = 0, \tag{2.8}$$
$$u(0) = 0 \quad \text{and} \quad W_p(1, u'(1)) - t_r = 0,$$

where W_p means differentiation of W with respect to the second variable. Note, that (2.7) has in general no global minimizer due to the fact, that W is nonconvex in u'. In particular, existence of a solution of (2.8) cannot be guaranteed by the direct methods of the calculus of variations.

We introduce the relaxed variational problem by adding an additional strain gradient term intended to model interfacial energy:

$$J_\varepsilon(u) := \int_0^1 \left(\frac{\varepsilon}{2} u''^2 + W(x, u') + B(u, x) \right) dx - t_r u(1), \tag{2.9}$$
$$u(0) = 0.$$

The Euler–Lagrange equation of equilibrium is the fourth order equation

$$-\varepsilon u^{(4)} + \frac{d}{dx} \left[W_p(x, u') \right] + b(u, x) = 0 \tag{2.10}$$

with boundary conditions

$$u(0) = u''(0) = u''(1) = 0, \qquad \varepsilon u'''(1) = W_p(1, u'(1)) - t_r. \tag{2.11}$$

Integration of (2.10) yields the system (with $\sigma(x, p) = W_p(x, p)$)

$$u' = z,$$
$$-\varepsilon z'' + \sigma(x, z) = \int_x^1 b(u(\tau), \tau) d\tau + t_r, \tag{2.12}$$
$$u(0) = z'(0) = z'(1) = 0.$$

Note, that (2.12) has certainly a solution for every $\varepsilon > 0$, namely the global minimizer obtained by the direct methods of the calculus of variations. But in general the Euler–Lagrange equation has a lot of solutions, in particular unstable solutions.

The plan is as follows: After giving the basic definition for a Young measure solution we prove some a–priori bounds on $(z_n)_{n\in\mathbb{N}}$ of solutions of (2.12) for corresponding $(\varepsilon_n)_{n\in\mathbb{N}}$ as $\varepsilon_n \searrow 0$. These bounds ensure that the sequence generates a Young measure ν. Moreover we prove that ν is a (generalized) solution of (2.8) in the sense described below. In particular, this implies regularity of the function $x \mapsto \langle \nu_x, \sigma \rangle$, where σ denotes the stress. This can be viewed as a generalization of the first Weierstrass–Erdmann corner condition. By obtaining similar estimates like the one in [43] especially involving the Young measure representation we can show, that the Young measure is a Dirac measure for almost every x, if the right hand side of (2.12)$_2$ is in the region where $p \mapsto \sigma(x, p)$ is injective.

We obtain the limit of the sequence $(u_n, z_n)_{n\in\mathbb{N}}$ of solutions of (2.12) in terms of a Young measure and prove, that this Young measure is a solution of (2.8) in the following sense:

Definition 5. A measurable map

$$\nu : [0, 1] \mapsto \text{Prob}(\mathbb{R})$$

is called a Young–measure solution of (2.8), if ν satisfies for almost every $x \in [0,1]$ the equations

$$
\begin{aligned}
\frac{d}{dx} \langle \nu_x, \sigma \rangle &= -b(u,x) + t_r, \\
\langle \nu_x, Id \rangle &= u'(x)
\end{aligned}
\tag{2.13}
$$

where $u \in W^{1,\infty}(0,1) \cap \{v \mid v(0) = 0\}$. In particular, $x \mapsto \langle \nu_x, \sigma \rangle \in C^1(0,1)$ because b was assumed to be continuous in every variable.

Remark 6. (i) Every ν satisfying (2.13) is also a solution in the sense of distributions, i.e.

$$
\int_0^1 \langle \nu_x, \sigma \rangle \, \varphi'(x) - b(u,x)\varphi(x)dx - t_r\varphi(1) = 0
\tag{2.14}
$$

for every $\varphi \in C^1(0,1) \cap \{v \mid v(0) = 0\}$ is valid.

(ii) If (2.8) has a solution u in a classical Sobolev–space involving the first derivative, then $\nu_x = \delta_{u'(x)}$ satisfies (2.13) for almost every $x \in [0,1]$. In particular, Definition 5 is an extension of the classical definition of a solution which fulfills the equation pointwise almost everywhere.

2.4   A–priori bounds

For the proofs of Theorem 7 and Lemma 8 we refer to the proofs in Chapter 3, Theorem 19, Lemma 22 and Lemma 23.

Theorem 7. *Let σ and b satisfy the following growth conditions:*

(i) *We assume for every $x \in [0,1]$*

$$
\lim_{p \to \pm\infty} \frac{\sigma(x,p)p}{||p||^{t+1}} \geq K(x)
\tag{2.15}
$$

with some $K(x) \geq 0$ and

$$
K(x) > 0 \text{ for every } x \in \tilde{\Omega},
\tag{2.16}
$$

where $\tilde{\Omega} \subseteq [0,1]$ is measurable with $|\tilde{\Omega}| > 0$.

(ii) *Let*

$$
||b(u,x)|| \leq c_3(x) \, ||u||^r + c_4(x)
\tag{2.17}
$$

with $r < t$ and $c_3(x), c_4(x) > 0$ for every $x \in [0,1]$.

Furthermore let $(z_n)_{n \in \mathbb{N}}$ be a sequence of solutions of (2.12) for corresponding $(\varepsilon_n)_{n \in \mathbb{N}}$ with $\varepsilon_n \searrow 0$. Then we deduce

$$
||z_n||_\infty \leq C
\tag{2.18}
$$

for every $n \in \mathbb{N}$.

Moreover we can state the following:

Lemma 8. *Let $(z_n)_{n \in \mathbb{N}}$ be a sequence like in Theorem 7. Then we obtain:*

(i) *The sequence $(\varepsilon_n z_n'')_{n \in \mathbb{N}}$ is uniformly bounded in $C^0([0,1])$.*

(ii) *The sequence $(\sqrt{\varepsilon_n} z_n')_{n \in \mathbb{N}}$ is uniformly bounded in $L^2(0,1)$.*

There exists a subsequence of $(z_n)_{n \in \mathbb{N}}$, not relabelled, such that the following holds:

(iii) *We have $\varepsilon_n z_n'' \rightharpoonup^* 0$ in $L^\infty(0,1)$ and*

(iv) *$\varepsilon_n z_n' \to 0$ in $C([0,1])$.*

2.5 Results

By virtue of Corollary 3 and Theorem 7 we obtain, that the sequence $(z_n)_{n \in \mathbb{N}}$ generates a Young measure ν satisfying

$$\langle \nu_x, Id \rangle = z(x) \quad \text{for almost every } x \in [0,1], \tag{2.19}$$

where z is the weak limit of $(z_n)_{n \in \mathbb{N}}$. Moreover we can prove the following:

Theorem 9. *The Young measure ν generated by an arbitrary sequence $(z_n)_{n \in \mathbb{N}}$ of solutions of (2.12) satisfies the equation*

$$\langle \nu_x, W_p(x, \cdot) \rangle = \int_{\mathbb{R}} W_p(x, \lambda) d\nu_x(\lambda) = \int_x^1 b(u(\tau), \tau) d\tau + t_r \tag{2.20}$$

for almost every $x \in [0,1]$, where $u \in W^{1,\infty}(0,1) \cap \{v \mid v(0) = 0\}$ is the limit of $(u_n)_{n \in \mathbb{N}}$. In particular, the function

$$x \mapsto \langle \nu_x, W_p(x, \cdot) \rangle \in C^1(0,1). \tag{2.21}$$

This can be viewed as a generalization of the first Weierstrass–Erdmann corner condition.
Moreover, ν is a Young measure solution by Definition 5.

Proof. We consider a sequence (u_n, z_n) of solutions of (2.12) for corresponding ε_n with $\varepsilon_n \searrow 0$. By virtue of Theorem 7 and Theorem 1 we deduce (after choosing a subsequence if necessary), that $(z_n)_{n \in \mathbb{N}}$ generates a Young measure ν. The compact embedding $C^1([0,1]) \hookrightarrow C^0([0,1])$ implies uniform convergence of $(u_n)_{n \in \mathbb{N}}$ to some $u \in C^0([0,1])$.
Furthermore we have for every $n \in \mathbb{N}$

$$-\varepsilon_n z_n'' + \sigma(x, z_n) = \int_x^1 b(u_n(\tau), \tau) d\tau + t_r.$$

By Corollary 3 we obtain

$$\sigma(x, z_n) \rightharpoonup^* \langle \nu_x, W_p(x, \lambda) \rangle \quad \text{in } L^\infty(0,1)$$

and moreover

$$\int_x^1 b(u_n(\tau), \tau) d\tau \to \int_x^1 b(u(\tau), \tau) d\tau$$

by the above mentioned uniform convergence. Taking Lemma 8 (iii) into account we end up with (2.20). □

Our next goal is to determine the Young measure in certain regions to be a Dirac measure. Following the idea of [43] the key ingredient to get such a result is the following inequality:

Lemma 10. *Let $\varphi \in C^1([0,1] \times \mathbb{R}, \mathbb{R})$ an arbitrary function with $\varphi_z(x,z) \geq 0$ for every $x \in [0,1]$. Then we deduce the inequality*

$$\int_0^1 \langle \nu_x, W_p(x,\lambda)\varphi(x,\lambda)\rangle \, dx \leq \int_0^1 \langle \nu_x, W_p(x,\lambda)\rangle \, \langle \nu_x, \varphi(x,\lambda)\rangle \, dx, \qquad (2.22)$$

where ν is the Young measure generated by a sequence of solutions of (2.12).

Proof. Multiplying (2.12) by φ and integrating over the unit interval yields

$$\int_0^1 (-\varepsilon z''(x)\varphi(x,z) + W_p(x,z)\varphi(x,z)) \, dx = \int_0^1 \left(\int_x^1 b(u(\tau),\tau) d\tau + t_r \right) \varphi(x,z) dx.$$

By integration by parts of the first term we obtain by taking the boundary conditions and the assumption on φ into account:

$$\int_0^1 W_p(x,z)\varphi(x,z) dx =$$
$$= \int_0^1 \left(-\varepsilon z' \varphi_x(x,z) - \varepsilon z'^2 \varphi_z(x,z) + \left(\int_x^1 b(u(\tau),\tau) d\tau + t_r \right) \varphi(x,z) \right) dx \leq$$
$$\leq \int_0^1 \left(-\varepsilon z' \varphi_x(x,z) + \left(\int_x^1 b(u(\tau),\tau) d\tau + t_r \right) \varphi(x,z) \right) dx.$$
$$(2.23)$$

Consider a sequence $(\varepsilon_n, z_n)_{n \in \mathbb{N}}$ with $\varepsilon_n \searrow 0$ and z_n is a solution of (2.12) for corresponding ε_n. Let ν be the Young measure generated by $(z_n)_{n \in \mathbb{N}}$. By Lemma 8 and (2.18) we obtain

$$\lim_{n \to \infty} \int_0^1 -\varepsilon z_n' \varphi_x(x, z_n(x)) dx = 0$$

and (2.23) implies

$$\int_0^1 \langle \nu_x, W_p(x,\lambda)\varphi(x,\lambda)\rangle \, dx \leq \int_0^1 \left[\left(\int_x^1 b(u(\tau),\tau) d\tau + t_r \right) \langle \nu_x, \varphi(x,\lambda)\rangle \right] dx.$$

Theorem 9 implies immediately (2.22). □

Similar to [43] we want to obtain an estimate which is more convenient for our forthcoming analysis. Therefore we consider the product measure $\tilde{\nu} := \nu \otimes \nu$ over $\mathbb{R} \times \mathbb{R}$.

Lemma 11. *Let φ be a test function as in Lemma 10 and let ν be a Young measure generated by an arbitrary sequence of solutions of (2.12). Then we get the inequality*

$$\int_0^1 \left[\int_{\mathbb{R}^2} \big(\varphi(x,\lambda) - \varphi(x,\tau)\big)\big(W_p(x,\lambda) - W_p(x,\tau)\big) d\big(\nu_x(\lambda) \otimes \nu_x(\tau)\big) \right] dx \leq 0. \quad (2.24)$$

Proof. Let $\tilde{\nu} := \nu \otimes \nu$ and let ν be the Young measure generated by a sequence of solutions of (2.12). Furthermore let $(\lambda, \tau) \in \mathbb{R}^2$ and φ an arbitrary test function as the one considered in Lemma 10, then we obtain by Fubini's theorem

$$\langle \nu_x, W_p \rangle \langle \nu_x, \varphi \rangle =$$
$$= \frac{1}{2} \left[\int_{\mathbb{R}} W_p(x,\lambda) d\nu_x(\lambda) \int_{\mathbb{R}} \varphi(x,\lambda) d\nu_x(\lambda) + \int_{\mathbb{R}} W_p(x,\tau) d\nu_x(\tau) \int_{\mathbb{R}} \varphi(x,\tau) d\nu_x(\tau) \right] =$$
$$= \frac{1}{2} \left[\int_{\mathbb{R}} W_p(x,\lambda) d\nu_x(\lambda) \int_{\mathbb{R}} \varphi(x,\tau) d\nu_x(\tau) + \int_{\mathbb{R}} W_p(x,\tau) d\nu_x(\tau) \int_{\mathbb{R}} \varphi(x,\lambda) d\nu_x(\lambda) \right] =$$
$$= \frac{1}{2} \left[\int_{\mathbb{R}^2} \big(W_p(x,\lambda)\varphi(x,\tau) + W_p(x,\tau)\varphi(x,\lambda)\big) d(\tilde{\nu}_x(\lambda,\tau)) \right]. \quad (2.25)$$

On the other hand we have due to the fact that ν_x is a probability measure for almost every $x \in [0,1]$

$$\int_{\mathbb{R}} W_p(x,\lambda)\varphi(x,\lambda) d\nu_x(\lambda) = \qquad\qquad (2.26)$$
$$= \frac{1}{2} \left[\int_{\mathbb{R}^2} \big(W_p(x,\lambda)\varphi(x,\lambda) + W_p(x,\tau)\varphi(x,\tau)\big) d(\tilde{\nu}_x(\lambda,\tau)) \right].$$

Equation (2.25) and (2.26) implies

$$\langle \nu_x, W_p\varphi \rangle - \langle \nu_x, W_p \rangle \langle \nu_x, \varphi \rangle =$$
$$= \frac{1}{2} \left[\int_{\mathbb{R}^2} W_p(x,\lambda)\varphi(x,\lambda) + W_p(x,\tau)\varphi(x,\tau) - \right.$$
$$\left. - W_p(x,\lambda)\varphi(x,\tau) - W_p(x,\tau)\varphi(x,\lambda) \, d(\tilde{\nu}_x(\lambda,\tau)) \right] =$$
$$= \frac{1}{2} \left[\int_{\mathbb{R}^2} \big(W_p(x,\lambda) - W_p(x,\tau)\big)\big(\varphi(x,\lambda) - \varphi(x,\tau)\big) d(\tilde{\nu}_x(\lambda,\tau)) \right]$$

and Lemma 10 gives the result. $\qquad\qquad\qquad\qquad\qquad\qquad\qquad\qquad\square$

We introduce certain points

$$r_a(x) := \min\{r \in \mathbb{R} \mid W_p(x,r) = W_p(x,c_2(x))\},$$
$$r_b(x) := \max\{r \in \mathbb{R} \mid W_p(x,r) = W_p(x,c_1(x))\},$$

where $c_1(x), c_2(x)$ are are defined by (2.6). Furthermore we assume

$$W_p \in C^1([0,1] \times \mathbb{R}, \mathbb{R}),$$

which implies immediately r_a, $r_b \in C^1([0,1])$. This is in fact a restriction to the inhomogenity of W. Note, that $W_p(x,\lambda)$ is injective for $\lambda \in (-\infty, r_a(x)) \cup (r_b(x), \infty)$. We define²

$$T(x) := \mathbb{R}\backslash[W_p(x,c_2(x)), W_p(x,c_1(x))]$$

and furthermore

$$A(u) := \left\{ x \in [0,1] \mid \int_x^1 b(s, u(s))ds + t_r \in T(x) \right\} \qquad (2.27)$$

for some $u \in C^0(0,1) \cap \{v \mid v(0) = 0\}$. Note, that A is an open subset of $[0,1]$ by the continuity of $x \mapsto \int_x^1 b(s, u(s))ds$ and by the assumptions on W. Now we can prove the following statement:

Theorem 12. *Let $(z_n)_{n \in \mathbb{N}}$ be a sequence of solutions of (2.12), let z be the weak limit of the sequence and let $u(x) = \int_0^x z(s)ds$. Furthermore let ν be the Young measure generated by this sequence. Then we obtain for almost every $x \in A(u)$*

$$\nu_x = \delta_{z(x)}.$$

In particular, by Corollary 4, there exists a subsequence, not relabelled, of $(z_n)_{n \in \mathbb{N}}$ such that $z_n(x) \to z(x)$ for almost every $x \in A(u)$.

Proof. Let $(z_n)_{n \in \mathbb{N}}$ be a sequence of solutions of (2.12) for $\varepsilon_n \searrow 0$ and let ν be the Young measure generated by this sequence. Let u be the limit of the corresponding sequence $(u_n)_{n \in \mathbb{N}}$.
To deduce the result we define a test function $\varphi \in C^1([0,1] \times \mathbb{R}, \mathbb{R})$ with the following properties:

(i) $\varphi(x, \cdot)_{|\mathbb{R} \setminus [r_a(x), r_b(x)]}$ is strictly monotonically increasing for every $x \in [0,1]$,

(ii) $\varphi(x, \cdot)_{|[r_a(x), r_b(x)]} \equiv$ const. for every $x \in [0,1]$.

Moreover we define

$$f(x, \lambda, \tau) := (W_p(x, \lambda) - W_p(x, \tau)) (\varphi(x, \lambda) - \varphi(x, \tau))$$
$$f : [0,1] \times \mathbb{R} \times \mathbb{R} \to \mathbb{R}$$

and by the properties of φ we obtain for every $x \in [0,1]$

$$f(x, \lambda, \tau) \begin{cases} > 0 & , \quad \text{if } (\lambda, \tau) \notin [r_a(x), r_b(x)]^2 \text{ and } \lambda \neq \tau \\ = 0 & , \quad \text{elsewhere.} \end{cases} \qquad (2.28)$$

By virtue of (2.24) and (2.28) we deduce

$$\int_{\mathbb{R}^2} f(x, \lambda, \tau) \, d(\nu_x(\lambda) \otimes \nu_x(\tau)) = 0 \text{ for almost every } x \in [0,1].$$

Hence we have

$$\text{supp}(\nu_x \otimes \nu_x) \subseteq \{(\lambda, \tau) \in \mathbb{R}^2 \mid f(x, \lambda, \tau) = 0\} \text{ for almost every } x \in [0,1]. \qquad (2.29)$$

Assume for a moment, that the support of ν_x consists of at least two different points $\{c(x), d(x)\}$, which implies

$$\{c(x), d(x)\} \times \{c(x), d(x)\} \in \text{supp}(\nu_x \otimes \nu_x).$$

Because of (2.28) and (2.29) we deduce $\{c(x), d(x)\} \in [r_a(x), r_b(x)]$. Furthermore the points c, d were chosen arbitrary and thus we obtain

$$\operatorname{supp} \nu_x \subseteq [r_a(x), r_b(x)], \tag{2.30}$$

which implies by Theorem 9

$$\int_x^1 b(s, u(s))ds + t_r = \langle \nu_x, W_p \rangle \in [W_p(x, c_2(x)), W_p(x, c_1(x))].$$

We conclude immediately $x \notin A(u)$. This yields the desired result. □

In the same way one can prove the following corollary just using $z(x) = \langle \nu_x, Id \rangle$ for a.e. $x \in [0, 1]$, where z is the weak limit of $(z_n)_{n \in \mathbb{N}}$ and ν is the Young measure generated by $(z_n)_{n \in \mathbb{N}}$.

Corollary 13. *Let $(z_n)_{n \in \mathbb{N}}$ and ν be as in Theorem 12 and let z be the weak limit of $(z_n)_{n \in \mathbb{N}}$. Then for almost every $x \in [0, 1]$ with $z(x) \in \mathbb{R} \backslash [r_a(x), r_b(x)]$ we obtain $\nu_x = \delta_{z(x)}$.*
Moreover we have: If $\operatorname{supp} \nu_x \subseteq (-\infty, r_a(x)) \cup (r_b(x), \infty)$ is valid, then we get also $\nu_x = \delta_{z(x)}$.

Another immediate consequence of (2.30) is the following:

Corollary 14. *Let the assumptions be as in Corollary 13 and define $u(x) := \int_0^x z(s)ds$. Then for almost every $x \in [0, 1] \setminus A(u)$ we have*

$$\operatorname{supp} \nu_x \subseteq [r_a(x), r_b(x)].$$

Proof. We assume some $x \in [0, 1] \setminus A(u)$, such that

$$c \notin [r_a(x), r_b(x)] \text{ and } \{c\} \in \operatorname{supp} \nu_x. \tag{2.31}$$

Without loss of generality we assume $c < r_a(x)$ and hence

$$\sigma(x, c) < \int_x^1 b(s, u(s))ds + t_r = \langle \nu_x, W_p \rangle \in [W_p(x, c_2(x)), W_p(x, c_1(x))].$$

In particular, $\operatorname{supp} \nu_x$ consists of at least two points $\{c, d\}$ with $d > c$. By the same procedure as in Theorem 12 we obtain (see (2.30))

$$\{c, d\} \subseteq [r_a(x), r_b(x)]$$

contradicting (2.31). □

2.6 Concluding remarks

The Young measure $\nu = (\nu_x)_{x \in \Omega}$ generated by a sequence $(v_n)_{n \in \mathbb{N}}$, where $v_n : \Omega \mapsto \mathbb{R}^N$, Ω denotes some open set, has the following interpretation: The measure ν_{x_0} can be thought of the limiting probability of values of $(v_n)_{n \in \mathbb{N}}$ in a small neighborhood of

$x_0 \in \Omega$. In order to be mathematically more precise we have the following: First we define

$$\mu_{x_0,\delta,v_n}(C) := \frac{|\{x \in B_\delta(x_0) \mid v_n(x) \in C\}|}{|B_\delta(x_0)|},$$

where C denotes some measurable set in $\mathbb{R}^N$ and $|\cdot|$ denotes the Lebesgue measure. Then we have for every $f \in C_0(\mathbb{R}^N)$

$$\lim_{\delta \searrow 0} \lim_{n \to \infty} \langle \mu_{x_0,\delta,v_n}, f \rangle = \langle \nu_{x_0}, f \rangle .$$

Hence we obtain by Theorem 12: Let $u(x) := \int_0^x z(s)ds$, where z is the weak limit of the generating sequence $(z_n)_{n \in \mathbb{N}}$. If $\int_x^1 b(u(\tau),\tau)d\tau + t_r \in A(u)$, where $A(u)$ is defined by (2.27), then the limit of the generating sequence $(z_n)_{n \in \mathbb{N}}$ is determined, because the resulting Young measure ν is a Dirac measure for almost every $x \in A(u)$. In particular, we have by Vitalis theorem

$$\lim_{n \to \infty} z_{n|A(u)} = z_{|A(u)} \quad \text{in } L^1(A).$$

On the other hand, if $x \notin A(u)$, then ν_x is not necessarily a Dirac measure which indicates the possibility of a strongly oscillatory behavior of the sequence $(z_n)_{n \in \mathbb{N}}$ in this region. However, by virtue of Corollary 14 the values attained by $(z_n)_{n \in \mathbb{N}}$ in $[0,1] \setminus A(u)$ are in the interval $[r_a(x) - \delta, r_b(x) + \delta]$ for arbitrary $\delta > 0$ and for $n = n(\delta)$ sufficiently large. Thus the amplitude of the possible oscillation is restricted by these two values.

Chapter 3

The continuation method approach

3.1 Introduction

One of the most powerful tools in nonlinear analysis is the Leray–Schauder degree. In nonlinear elasticity the degree was used to obtain solutions of the Euler–Lagrange equation corresponding to nonconvex variational problems. Especially in problems in which the solution is a real valued function one can apply the classical degree typically to the Euler–Lagrange equation of the singular perturbed problem.

In [27] the existence of classical solutions of the Euler–Lagrange equation was proved corresponding to the problem

$$E(u) = \int_0^1 W(u') + G(u) \, dx,$$

with Dirichlet boundary conditions, where W is a double–well potential and G is typically (but not necessarily) concave. Furthermore G is assumed to have an isolated maximum at $u = 0$. Using global bifurcation analysis for the variational problem

$$E_\varepsilon(\lambda, u) = \int_0^1 \left(\frac{\varepsilon}{2} u''^2 + W(\lambda + u') + G(u) \right) dx,$$

where λ is the bifurcation parameter, one obtains nontrivial critical points of $E_\varepsilon(0, u)$ with certain patterns. These patterns allow one to prove existence of the singular limit in a classical function space as $\varepsilon \searrow 0$. Thus one obtains infinitely many nontrivial critical points of $E_0(0, u)$ with certain symmetries. Under additional assumptions on W and G it is also possible to show uniqueness of the global minimizers in the obtained symmetry classes.

In [25] and [26] the Cahn–Hilliard energy functional $E_0(u)$ is considered over the unit square under the constraint of constant mass m. The perturbed functional is defined by

$$E_\varepsilon(u) = \int_{[0,1]^2} \left(\frac{\varepsilon}{2} \|\nabla u\|^2 + W(u) \right) dx$$

and the constraint by

$$\int_{[0,1]^2} u \, dx = m.$$

The goal in [26] is to select (physically) preferred patterns of critical points, that means solutions of the Euler–Lagrange equation with minimal interface. These solutions are obtained by a global bifurcation analysis of the EL equation of the singular perturbed variational problem E_ε using the mass m as the bifurcation parameter and considering the singular limit. The existence of this limit in a classical function space is guaranteed by the location of maxima and minima of the critical points of E_ε, which are fixed for all solutions on global branches.

The key idea in both cases (and also in this one under consideration) is to figure out certain properties of a solution of the perturbed EL equation which enables one to obtain a singular limit result in appropriate function spaces.

Another approach for the one–dimensional elastic bar involving bifurcation analysis can be found in [42], where the effect of the spinodal region on the structure of equilibria is studied. Under consideration are the two parabola model in which the spinodal region reduces to one point, the three parabola model, where the stress is piecewise affine but the spinodal region contains an interval and last the usual smooth nonconvex energy density. Also in this approach the bifurcation analysis is applied to the perturbed variational problem and it enables them to study the evolution of branches of local minima.

In three–dimensional elasticity problems global continuation results were obtained in [17], [19] and [20] by employing generalized degree theoretic methods. In this case the resulting solution continua are characterized not only by the usual two alternatives given in [37], but also by a third one concerning the termination of the branch. This phenomenon can happen due to loss of local injectivity, elipticity and/or the complementing condition.

In this work we generalize the result given in [18] for a one–dimensional elastic bar placed in a soft loading device with (in general live) body–force $-\frac{\partial B}{\partial u} = b \geq 0$. The total energy of the bar is defined by

$$J(\lambda, u) := \int_0^1 \left(W(u') + B(\lambda, u, x) \right) dx,$$
$$u(0) = 0. \tag{3.1}$$

As in [18] we use the global continuation method to obtain a global branch of solutions of the EL equation of the following singular perturbed problem:

$$J_\varepsilon(\lambda, u) := \int_0^1 \left(\frac{\varepsilon}{2} u''^2 + W(u') + B(\lambda, u, x) \right) dx,$$
$$u(0) = 0.$$

In this approach we figure out some properties of the solutions via continuum methods to deduce that the sequence of deformations is relative compact in $L^1(0,1)$ as $\varepsilon \searrow 0$. The outline of the work is as follows: In section 1 we introduce the variational problem under consideration and the corresponding EL equations. Furthermore we state the assumptions on the potentials.

A global analysis result will be given in section 2 and by considering a so called one–well potential W we rule out that the branch can form a loop. Due to some physically reasonable growth conditions we prove the unboundedness of the branch for arbitrary $\varepsilon > 0$ especially in the λ–direction. This procedure is one of the main differences to

[18], where a similar result is obtained by the maximum principle which does not apply to the more general body force considered in this work. Our approach applies also in [18]. Thus all the results stated there remain valid if one replaces the condition on the body force

$$b(\lambda, u(x), x) \leq \tilde{b}(\lambda, x)$$

for every $u \in \{y \in C^0(0,1) \mid y(0) = 0\}$, $x \in [0,1]$, by the more general growth conditions stated in Theorem 19. But note, that we have to impose a condition on W which is not necessary in [18].

The third section is the core of the paper. We consider a sequence $(u_n)_{n \in \mathbb{N}}$ of solutions of the EL equation (3.5) for corresponding ε_n as $\varepsilon_n \searrow 0$. First we prove some a–priori bounds on the sequence based upon the EL equation. Furthermore by degree theoretic methods we obtain under the assumption that the body–force changes sign from "+" to "−" a subsequence of $(u_n)_{n \in \mathbb{N}}$ relatively compact in $W^{1,1}(0,1)$. In particular, the sequence $(u'_n)_{n \in \mathbb{N}}$ converges pointwise in $(0,1)$ and moreover the limit satisfies the first and second Weierstrass–Erdmann corner condition and solves also the EL equation for the unperturbed variational problem (3.1). To obtain such a result we especially use the well–known maximum principle. But in contrast to [18] we have additionally to employ some techniques based on monotonicity of u'_n. Unfortunately this rules out to get similar results for the double–well potential as in [18]. In order to state analogues results by employing the same techniques for more general body forces we have to restrict ourselves to the dead load case with special symmetries.

In the fourth section we use a result by Alexander (see [1] for details) to prove that the pointwise limit also forms a global continuum unbounded in the λ–direction.

By considering the dead–load case we show in the last section that the limit obtained in section 3 is actually the global minimizer of the unperturbed problem. We remark, that the results of section 4 and 5 are proved in the same way as in [18].

3.2   Global analysis

We consider the relaxed variational problem

$$J_\varepsilon(\lambda, u) := \int_0^1 \left(\frac{\varepsilon}{2} u''^2 + W(u') + B(\lambda, u, x) \right) dx, \qquad (3.2)$$
$$u(0) = 0.$$

The Euler–Lagrange equation of equilibrium is the fourth order equation

$$-\varepsilon u^{(4)} + \frac{d}{dx} \left[W'(u') \right] + b(\lambda, u, x) = 0 \qquad (3.3)$$

with boundary conditions

$$u(0) = u''(0) = u''(1) = 0, \qquad \varepsilon u'''(1) = W'(u'(1)). \qquad (3.4)$$

Integration of (3.3) yields the system with $\sigma = W'$

$$
\begin{aligned}
u' &= z, \\
-\varepsilon z'' + \sigma(z) &= \int_x^1 b(\lambda, u(\tau), \tau)d\tau, \\
u(0) &= z'(0) = z'(1) = 0, \\
\int_0^1 \sigma(z(x))dx &= \int_0^1 \left[\int_x^1 b(\lambda, u(\tau), \tau)d\tau\right] dx.
\end{aligned}
\tag{3.5}
$$

We assume $b \in C^0(\mathbb{R} \times \mathbb{R} \times [0,1], \mathbb{R})$ and

$$
b(0, \cdot, \cdot) \equiv 0. \tag{3.6}
$$

For the sequel of the chapter we consider a one–well potential W with associated $\sigma := W'$. Here we impose the following assumptions on $W \in C^2(\mathbb{R})$: There are numbers $0 < \nu_1 < \nu_2$ such that

$$
\begin{aligned}
&W(\nu) \geq W(0) = 0 \text{ for every } \nu \in \mathbb{R}, \\
&\lim_{\nu \to \pm\infty} W(\nu) = \infty, \\
&W'(0) = 0, \\
&W'(\nu) \begin{cases} < 0, & \text{for } \nu \in (-\infty, 0) \\ > 0, & \text{for } \nu \in (0, \infty), \end{cases} \\
&W''(\nu) \begin{cases} < 0, & \text{for } \nu \in (\nu_1, \nu_2) \\ > 0, & \text{otherwise.} \end{cases}
\end{aligned}
\tag{3.7}
$$

Moreover we define

$$
\begin{aligned}
\nu_a &:= \min\left\{\nu \in \mathbb{R} \mid W'(\nu) = W'(\nu_2)\right\}, \\
\nu_b &:= \max\left\{\nu \in \mathbb{R} \mid W'(\nu) = W'(\nu_1)\right\}.
\end{aligned}
\tag{3.8}
$$

To obtain results by the continuation method we have to convert (3.5) into an operator form equation. Therefore we follow exactly the lines of [18]. Nevertheless we state it in detail for the reader's convenience. The only difference is that we have to obtain more regularity to proceed our analysis.

First we define a linear operator S by

$$
w := Sf
$$

and w is the unique solution of

$$
w' = f, \, 0 < x < 1, \, w(0) = 0
$$

for $f \in C^0([0,1])$. Furthermore we define an operator T by

$$
v = Tg
$$

and v is the unique solution of

$$v'' = g,\, 0 < x < 1,\, v'(0) = v'(1) = 0,$$

$$\int_0^1 v\, dx = 0 \text{ for } g \in \left\{ y \in C^0([0,1]) \mid \int_0^1 y dx = 0 \right\}.$$

We define the following function spaces and let $\tilde{Y} := C^0([0,1])$:

$$\tilde{Y}_0 := \left\{ y \in \tilde{Y} \mid y(0) = 0 \right\}, \tilde{Y}_1 := \left\{ y \in \tilde{Y} \mid \int_0^1 y dx = 0 \right\}$$

and

$$X := \left\{ y \in C^3([0,1]) \mid y(0) = 0 \right\}, Z := \left\{ y \in C^4([0,1]) \mid y'(0) = y'(1) = 0, \int_0^1 y dx = 0 \right\}.$$

Each function space is equipped with the usual supremum norm to obtain a Banach space. Furthermore we define $Y_j := \tilde{Y}_j \cap C^2([0,1])$ for $j = 0,1$ also endowed with the usual norm. Then it is easy to see that $S : Y \to X$ and $T : Y_1 \to Z$ are bounded operators.

Furthermore we define

$$\mu := \int_0^1 z(s)ds,\, v := z - \mu.$$

Note, that by the boundary condition $u(0) = 0$ we obtain $\mu = u(1)$.

Finally we define the triple

$$w := (u, v, \mu) \in \mathcal{W} \equiv Y_0 \times Y_1 \times \mathbb{R}$$

and $\mathcal{W}$ is endowed with the norm $||w||_{\mathcal{W}} = ||u||_{C^2} + ||v||_{C^2} + |\mu|$ and in particular $\mathcal{W}$ is a Banach space. By an easy calculation one obtains that (3.5) is equivalent to

$$w - H_\varepsilon(\lambda, w) = 0, \tag{3.9}$$

where $H_\varepsilon : \mathbb{R} \times \mathcal{W} \to \mathcal{W}$ is defined by

$$H_\varepsilon(\lambda, w) := \left(S(v + \mu), \frac{1}{\varepsilon} T \left[\sigma(v + \mu) \right.\right.$$

$$- \int_x^1 b(\lambda, u(\tau), \tau)d\tau - \int_0^1 \left(\sigma(v + \mu) - \int_x^1 b(\lambda, u(\tau), \tau)d\tau \right) dx \right], \tag{3.10}$$

$$\left. \int_0^1 \left(\sigma(v + \mu) - \int_x^1 b(\lambda, u(\tau), \tau)d\tau \right) dx + \mu \right).$$

Due to the compact embedding $X \hookrightarrow Y_0$ and $Z \hookrightarrow Y_1$ the operators $S : Y \to Y_0$ and $T : Y_1 \to Y_1$ are compact and it readily follows that $w \mapsto H_\varepsilon(\lambda, w)$ is also compact.

Moreover it is straightforward to verify that any solution (λ, w) of (3.9) delivers a classical solution $(\lambda, u, z) = (\lambda, u, v + \mu)$ of (3.5) and a classical solution (λ, u) of (3.3). With respect to (3.6) and (3.7) we obtain $H_\varepsilon(0,0) = 0$ and in particular $(\lambda = 0, u \equiv 0, z \equiv 0)$ is a solution of (3.5). The Frechet derivative of $w \mapsto H_\varepsilon(\lambda, w)$ at $(0,0)$ is given by

$$D_w H_\varepsilon(0,0)[h] = \left(S(h_2 + h_3), \frac{1}{\varepsilon} W''(0) T h_2, (1 + W''(0)) h_3 \right)$$

for all $h = (h_1, h_2, h_3) \in Y_0 \times Y_1 \times \mathbb{R}$.

Proposition 15. *The map* $I - D_w H_\varepsilon(0,0) : \mathcal{W} \to \mathcal{W}$ *is injective and thus bijective by the Riesz–Schauder theory. ("I" denotes the identity map).*

Proof. Let $h = (h_1, h_2, h_3) \in N(I - D_w H_\varepsilon(0,0))$, which is equivalent to

$$
\begin{aligned}
h_1 - S(h_2 + h_3) &= 0, \\
h_2 - \frac{1}{\varepsilon} W''(0) T h_2 &= 0, \\
-W''(0) h_3 &= 0.
\end{aligned} \qquad (3.11)
$$

We have to show $h \equiv 0$. Due to $W''(0) > 0$ we get from $(3.11)_3$ immediately $h_3 = 0$. Furthermore we obtain by $(3.11)_2$ that h_2 is a solution of

$$
\frac{\varepsilon}{W''(0)} h_2'' = h_2, \; h_2'(0) = h_2'(1) = 0, \; \int_0^1 h_2 \, dx = 0.
$$

Because $\frac{\varepsilon}{W''(0)} > 0$ we deduce $h_2 \equiv 0$. By equation $(3.11)_1$ we get $h_1 = S(0) \equiv 0$, which proves the assertion. $\qquad \square$

By the implicit function theorem we obtain the following :

Proposition 16. *Equation* (3.9) *has a local curve of solutions*

$$
(\lambda, w) = (\lambda, \tilde{w}(\lambda)), \quad |\lambda| < \delta, \qquad (3.12)
$$

where $\tilde{w}(0) = 0$. *Moreover,* (3.12) *yields all solutions of* (3.9) *in a sufficiently small neighborhood of* $(0,0)$.

The Leray–Schauder degree is well defined for $w \mapsto w - H_\varepsilon(\lambda, w)$ and from Proposition 16 we have

$$
i(I - H_\varepsilon(0, \cdot), w(\lambda)) = \pm 1 \text{ for } \lambda \in (-\delta, \delta)
$$

with $\delta > 0$ sufficiently small. A well known argument proved in [37] (see also [24]) employing the homotopy invariance of the degree yields the so called Rabinowitz alternatives:

Proposition 17. *Equation* (3.9) *admits a global branch of solution continua, denoted by* $\mathcal{C}_\varepsilon \subset \mathbb{R} \times \mathcal{W}$, *containing the local curve* (3.12) *and characterized by at least one of the following alternatives:*

(i) $\mathcal{C}_\varepsilon$ *is unbounded in* $\mathbb{R} \times \mathcal{W}$,

(ii) $\mathcal{C}_\varepsilon \backslash \{(0,0)\}$ *is connected.*

We remark, that this result was given in [18] by the same arguments. The only difference is, that by the above mentioned construction of the operators S and T it turns out, that $\mathcal{C}_\varepsilon$ is connected with respect to the C^2–topology. In [18] just a continuum in the C^1–topology is required.

3.3 Properties of the continuum

Our first aim is to rule out alternative (ii) of Proposition 17: A necessary condition for a loop is obviously the existence of a nontrivial solution of

$$\varepsilon z'' - \sigma(z) = 0, \quad 0 < x < 1$$
$$z'(0) = z'(1) = 0. \tag{3.13}$$

By the assumptions on W a standard phase–plane analysis like in [9] shows that $z \equiv 0$ is the only solution of (3.13) and accordingly we get:

Proposition 18. *The global continuum $\mathcal{C}_\varepsilon$ of Proposition 17 is characterized solely by alternative (i), i.e., $\mathcal{C}_\varepsilon$ is unbounded. Moreover, if we define*

$$\mathcal{C}_\varepsilon^{+(-)} = component\ of\ \mathcal{C}_\varepsilon \setminus \{(0,0)\},$$
$$containing\ \{(\lambda, \tilde{w}(\lambda)) \mid 0 < \lambda < \delta (-\delta < \lambda < 0)\}, \tag{3.14}$$

we then have the disjoint union $\mathcal{C}_\varepsilon = \mathcal{C}_\varepsilon^+ \cup \{(0,0)\} \cup \mathcal{C}_\varepsilon^-$, where $\mathcal{C}_\varepsilon^+$ and $\mathcal{C}_\varepsilon^-$ are each unbounded in $\mathbb{R} \times \mathcal{W}$.

Our next goal is to prove under physically reasonable growth conditions on W and the body force b, that the path projected to the λ–axis is $\mathbb{R}$.

Theorem 19. *Let σ and b satisfy the following growth conditions:*

(i) *We assume*

$$\|\sigma(\nu)\| \le c_1 + c_2 \|\nu\|^p \tag{3.15}$$

with constants c_1, $c_2 > 0$ and $p \ge 1$.

(ii) *Furthermore we assume*

$$\lim_{\nu \to \pm\infty} \frac{\sigma(\nu)\nu}{\|\nu\|^{p+1}} \ge K > 0 \tag{3.16}$$

and

(iii)

$$\sup_{\lambda \in [0,\hat{\lambda}]} \|b(\lambda, u, x)\| \le c_3 \|u\|^r + c_4 \tag{3.17}$$

for some $r < p$, $c_3 = c_3(\hat{\lambda})$, $c_4 = c_4(\hat{\lambda}) \ge 0$. Moreover we have c_3, $c_4 < \infty$ provided $\hat{\lambda} < \infty$ and both constants can be chosen independently of x.

Then the projection of $\mathcal{C}_\varepsilon^+$ on the λ–axis is $[0,\infty]$.

Proof. We consider $\mathcal{W} \equiv Y_0 \times Y_1 \times \mathbb{R}$ (with $Y_i := \tilde{Y}_i \cap C^2([0,1])$) and $\mathcal{W}$ is endowed with the norm $\|w\|_{\mathcal{W}} := \|u\|_{C^2} + \|v\|_{C^2} + |\mu|$.
We proof by contradiction: Suppose for an arbitrary but fixed $\varepsilon > 0$ the existence of $\lambda_0 < \infty$ such that

$$\mathcal{C}_\varepsilon^+ \subseteq \{(w, \lambda) \mid 0 < \lambda < \lambda_0 , w \in \mathcal{W}\}. \tag{3.18}$$

Because $\mathcal{C}_\varepsilon^+$ is unbounded (3.18) implies the existence of a sequence $(w_n, \lambda_n) \in \mathcal{W} \times \mathbb{R}$ such that

$$\lim_{n \to \infty} (\|w_n\|_{\mathcal{W}} + |\lambda_n|_{\mathbb{R}}) = \infty , \ |\lambda_n|_{\mathbb{R}} < \lambda_0$$

and in particular we have

$$\lim_{n\to\infty} (||u_n||_{C^2} + ||v_n||_{C^2} + |\mu_n|) = \infty. \tag{3.19}$$

We separate the proof into several steps:

Step 1   : We prove, that (3.19) implies

$$\lim_{n\to\infty} (||u_n||_{C^0} + ||v_n||_{C^0} + |\mu_n|) = \infty. \tag{3.20}$$

We assume $||u_n||_{C^0} + ||v_n||_{C^0} + |\mu_n| \le c$ for every $n \in \mathbb{N}$ with a certain constant $c > 0$. This implies

$$C \ge ||v_n||_{C^0} + |\mu_n| \ge \left| \, ||z_n||_{C^0} - |\mu_n| \, \right| + |\mu_n| \ge ||z_n||_{C^0} - |\mu_n| + |\mu_n| = ||z_n||_{C^0}$$

and the Euler–Lagrange equation yields due to our assumptions

$$||z_n''||_{C^0} \le \frac{1}{\varepsilon} \left(\left\| \sigma(z_n) - \int_x^1 b(\lambda_n, u_n(\tau), \tau) d\tau \right\|_{C^0} \right) \le$$

$$\le \frac{1}{\varepsilon} \left(||\sigma(z_n)||_{C^0} + \int_0^1 ||b(\lambda_n, u_n(\tau), \tau)|| \, d\tau \right) \le K.$$

Standard interpolation implies $||z_n'||_{C^0} \le M_\varepsilon ||z_n||_{C^0} + \varepsilon ||z_n''||_{C^0} \le K$ and thus we conclude $||z_n||_{C^2}$ and $||u_n||_{C^2}$ are uniformly bounded. By virtue of (3.19) we get $C \ge ||u_n||_{C^2} \ge |u_n(1)| = |\mu_n| \to \infty$ for $n \to \infty$, thus a contradiction.

Step 2   : We prove

$$\lim_{n\to\infty} ||z_n||_{L^{p+1}} = \infty. \tag{3.21}$$

By the fundamental theorem in calculus and by the identity $\mu = \int_0^1 z(s) ds$ we obtain

$$||u||_\infty + ||v||_\infty + |\mu| \le 4 ||z||_\infty. \tag{3.22}$$

Hence step 1 yields

$$\lim_{n\to\infty} ||z_n||_\infty = \infty. \tag{3.23}$$

We assume for a moment $||z_n||_{L^{p+1}} \le C$ for every $n \in \mathbb{N}$. Due to $||u_n||_{L^{p+1}} \le ||z_n||_{L^{p+1}}$ we get $(u_n)_{n\in\mathbb{N}}$ is uniformly bounded in $W^{1,p+1}(0,1)$ and also in $C^0(0,1)$ by continuous imbedding. The triple (u_n, z_n, λ_n) solves the equation

$$-\varepsilon z_n''(x) + \sigma(z_n(x)) = \int_x^1 b(\lambda_n, u_n(\tau), \tau) d\tau \tag{3.24}$$

for every $x \in [0,1]$. Multiplying (3.24) by z_n and integration over $(0,1)$ yields due to integration by parts and the Neumann boundary conditions

$$\int_0^1 \varepsilon z_n'(x)^2 dx = \int_0^1 \left[-\sigma(z_n(x))z_n(x) + \int_x^1 b(\lambda_n, u_n(\tau), \tau) d\tau z_n(x) \right] dx. \tag{3.25}$$

For the right hand side of (3.25) we obtain by virtue of Hölder's inequality (we denote by q the conjugate exponent of $p+1$) and (3.15) constants M_1, M_2, such that

$$\int_0^1 \varepsilon z_n'(x)^2 dx \le \int_0^1 \left(M_1 \left\| z_n \right\|^{p+1} + M_2 \right) dx + \left\| \int_x^1 b(\lambda_n, u_n(\tau), \tau) d\tau \right\|_{L^q} \left\| z_n \right\|_{L^{p+1}}. \tag{3.26}$$

Because of (3.17) and $\left\| u_n \right\|_\infty < C$ for every $n \in \mathbb{N}$ we obtain

$$\left(\left\| \int_x^1 b(\lambda_n, u_n(\tau), \tau) d\tau \right\|_{L^q} \right)_{n \in \mathbb{N}} \quad \text{is uniformly bounded}$$

and (3.26) implies therefore

$$\int_0^1 z_n'^2 dx \le K \quad \text{for all } n \in \mathbb{N}.$$

In particular $(u_n)_{n \in \mathbb{N}}$ is uniformly bounded in $H^2(0,1)$ (recall, that $p \ge 1$) and by continuous imbedding $H^2(0,1) \hookrightarrow C^1[0,1]$ we obtain

$$\left\| z_n \right\|_\infty \le \left\| z_n \right\|_{C^1} \le K \left\| u_n \right\|_{H^2} \le C,$$

a contradiction to (3.23).

Step 3 : Multiplying (3.25) with $\frac{1}{\left\| z_n \right\|_{L^{p+1}}^{p+1}}$ we obtain for every $n \in \mathbb{N}$

$$\frac{1}{\left\| z_n \right\|_{L^{p+1}}^{p+1}} \int_0^1 \varepsilon z_n'(x)^2 dx + \frac{1}{\left\| z_n \right\|_{L^{p+1}}^{p+1}} \int_0^1 \sigma(z_n(x)) z_n(x) dx =$$
$$= \frac{1}{\left\| z_n \right\|_{L^{p+1}}^{p+1}} \int_0^1 \left(\int_x^1 b(\lambda_n, u_n(\tau), \tau) d\tau z_n(x) \right) dx. \tag{3.27}$$

We claim:

(a) The right hand side of (3.27) converges to 0 for $n \to \infty$:
   By assumption (3.17) we obtain

$$\left\| \int_0^1 \left(\int_x^1 b(\lambda_n, u_n(\tau), \tau) d\tau z_n(x) \right) dx \right\| \le$$
$$\le \int_0^1 \left(\int_x^1 \left\| b(\lambda_n, u_n(\tau), \tau) \right\| d\tau \right) \left\| z_n(x) \right\| dx \le$$
$$\le \int_0^1 \left\| b(\lambda_n, u_n(\tau), \tau) \right\| d\tau \int_0^1 \left\| z_n(x) \right\| dx \le$$
$$\le \int_0^1 \left(c_3 \left\| u_n \right\|^r + c_4 \right) d\tau \left\| z_n \right\|_{L^1}.$$

Due to the continuous imbedding $L^{p+1}(0,1) \hookrightarrow L^r(0,1)$ and $\left\| u_n \right\|_{L^r} \le \left\| z_n \right\|_{L^r}$ the following holds:

$$\int_0^1 \left(c_3 \left\| u_n \right\|^r + c_4 \right) d\tau \left\| z_n \right\|_{L^1} \le c_3 \left\| z_n \right\|_{L^r}^r \left\| z_n \right\|_{L^1} + c_4 \left\| z_n \right\|_{L^1} \le$$
$$\le c_3 \left\| z_n \right\|_{L^{p+1}}^r \left\| z_n \right\|_{L^{p+1}} + c_4 \left\| z_n \right\|_{L^{p+1}} = c_3 \left\| z_n \right\|_{L^{p+1}}^{r+1} + c_4 \left\| z_n \right\|_{L^{p+1}}.$$

Because of $r < p$ and (3.21) assertion (a) is proved.

(b) We have

$$\lim_{n\to\infty} \frac{1}{||z_n||_{L^{p+1}}^{p+1}} \int_0^1 \sigma(z_n(x))z_n(x)dx \geq K > 0 \qquad (3.28)$$

with K given in (3.16).

Let $\delta > 0$ and (3.16) implies the existence of $\nu_u = \nu_u(\delta)$ and $\nu_o = \nu_o(\delta)$ such that

$$\frac{\sigma(\nu)\nu}{||\nu||^{p+1}} - K > -\frac{\delta}{2} \qquad (3.29)$$

for every $\nu < \nu_u$ and $\nu > \nu_o$. Multiplying (3.29) by $||\nu||^{p+1}$ one obtains

$$\sigma(\nu)\nu - K||\nu||^{p+1} > -\frac{\delta}{2}||\nu||^{p+1} \qquad (3.30)$$

for every $\nu < \nu_u$ and $\nu > \nu_o$. Due to (3.30) we get for an arbitrary $n \in \mathbb{N}$ and for all $x \in [0,1]$ with either $z_n(x) < \nu_u$ or $z_n(x) > \nu_o$ the estimate

$$\frac{1}{||z_n||_{L^{p+1}}^{p+1}} \left(\sigma(z_n(x))z_n(x) - K||z_n(x)||^{p+1} \right) > -\frac{\delta}{2} \frac{||z_n(x)||^{p+1}}{||z_n||_{L^{p+1}}^{p+1}}. \qquad (3.31)$$

Furthermore we have because of (3.15) for every $x \in [0,1]$ with $z_n(x) \in [\nu_u, \nu_o]$

$$\frac{1}{||z_n||_{L^{p+1}}^{p+1}} \left| \left| \sigma(z_n(x))z_n(x) - K||z_n(x)||^{p+1} \right| \right|$$

$$\leq \frac{1}{||z_n||_{L^{p+1}}^{p+1}} \left(||\sigma(z_n(x))z_n(x)|| + K||z_n(x)||^{p+1} \right) \leq$$

$$\leq \frac{1}{||z_n||_{L^{p+1}}^{p+1}} \left(c_2||z_n(x)||^{p+1} + c_1||z_n(x)|| + K||z_n(x)||^{p+1} \right) \leq$$

$$\leq \frac{1}{||z_n||_{L^{p+1}}^{p+1}} (\tilde{C} \max\{|\nu_u|, |\nu_o|\}^{p+1}) \leq \frac{1}{||z_n||_{L^{p+1}}^{p+1}} C.$$

Note, that the constant C just depends on δ. By (3.21) we deduce the existence of $n_1 = n_1(\delta)$, such that for every $n > n_1$ we obtain

$$\frac{1}{||z_n||_{L^{p+1}}^{p+1}} \left(\sigma(z_n(x))z_n(x) - K||z_n(x)||^{p+1} \right) > -\frac{\delta}{2} \qquad (3.32)$$

for every $x \in [0,1]$ such that $z_n(x) \in [\nu_u, \nu_o]$. We define

$$A_n := \{x \in [0,1] \mid z_n(x) < \nu_u \text{ or } z_n(x) > \nu_o\}.$$

Then we obtain by virtue of (3.31) and (3.32)

$$\frac{1}{||z_n||_{L^{p+1}}^{p+1}} \int_0^1 \sigma(z_n(x))z_n(x)dx - K =$$

$$= \frac{1}{||z_n||_{L^{p+1}}^{p+1}} \int_0^1 \left(\sigma(z_n(x))z_n(x) - K||z_n(x)||^{p+1} \right) dx =$$

$$= \frac{1}{||z_n||_{L^{p+1}}^{p+1}} \left[\left(\int_{[0,1]\cap A_n} + \int_{[0,1]\cap A_n^c} \right) \left(\sigma(z_n(x))z_n(x) - K||z_n(x)||^{p+1} \right) \right] \geq$$

$$\geq -\frac{\delta}{2} \frac{1}{||z_n||_{L^{p+1}}^{p+1}} ||z_n||_{L^{p+1}}^{p+1} - \frac{\delta}{2} = -\delta. \qquad (3.33)$$

Because $\delta > 0$ was chosen arbitrary, (3.28) holds.

By virtue of (3.27) we get

$$0 < K \leq \lim_{n \to \infty} \left(\frac{1}{||z_n||_{L^{p+1}}^{p+1}} \int_0^1 \varepsilon z_n'(x)^2 dx + \frac{1}{||z_n||_{L^{p+1}}^{p+1}} \int_0^1 \sigma(z_n(x)) z_n(x) dx \right) =$$
$$= \lim_{n \to \infty} \left(\frac{1}{||z_n||_{L^{p+1}}^{p+1}} \int_0^1 \left(\int_x^1 b(\lambda_n, u_n(\tau), \tau) d\tau z_n(x) \right) dx \right) = 0, \tag{3.34}$$

which is obviously a contradiction and the proof is done.   □

The above proof yields also a result, which is a bit stronger than the one stated in Theorem 19:

Corollary 20. *We consider the same growth conditions for σ as in Theorem 19 and furthermore we assume the existence of $\hat{\lambda} \leq \infty$, such that for all $\lambda \in [0, \hat{\lambda})$ and for every $x \in [0, 1]$ the following estimate is valid for a constant c and for the same K as in (3.16):*

$$||b(\lambda, u, x)|| < K ||u||^p + c.$$

Then we have a pair $(w, \lambda) \in \mathcal{W} \times \mathbb{R}$ such that $(w, \lambda) \in \mathcal{C}_\varepsilon^+$.

We omit the proof, because it is similar to the one of Theorem 19.

3.4 Some a–priori bounds

In the sequel of the chapter we will always assume the growth conditions given in Theorem 19. In particular we know for every $\lambda \in \mathbb{R}_0^+$ the existence of a classical solution of the system

$$u' = z,$$
$$- \varepsilon z'' + \sigma(z) = \int_x^1 b(\lambda, u(\tau), \tau) d\tau, \qquad 0 < x < 1 \tag{3.35}$$
$$u(0) = z'(0) = z'(1) = 0.$$

We consider a sequence $(\varepsilon_n, z_n)_{n \in \mathbb{N}}$, $\varepsilon_n \searrow 0$, and z_n is a solution of (3.35) for corresponding ε_n and for fixed $\lambda \in \mathbb{R}$. Our goal is to answer the question, whether $(z_n)_{n \in \mathbb{N}}$ converges to a function z in some appropriate function–space. To achieve this we first have to prove some a–priori bounds on the sequence.

Lemma 21. *Let σ and b satisfy the growth conditions of Theorem 19 and let $(z_n)_{n \in \mathbb{N}}$ be a sequence of solutions of (3.35) for corresponding ε_n. Then there exists a constant $C > 0$ with*

$$||z_n||_{L^{p+1}} \leq C \tag{3.36}$$

uniformly in $n \in \mathbb{N}$.

Proof. This statement is actually just a simple consequence of the proof of Theorem 19.

We consider for fixed $\lambda > 0$ a sequence $(\varepsilon_n, z_n)_{n \in \mathbb{N}}$ with the above mentioned properties and we assume $\lim_{n \to \infty} ||z_n||_{L^{p+1}} = \infty$. Completely analogous to step 3 in the proof of Theorem 19 we get due to (3.27) a contradiction and the proof is done.   □

By considering the Euler–Lagrange equation we obtain a much stronger bound which turns out to be very useful in our forthcoming analysis.

Lemma 22. *Let $(z_n)_{n \in \mathbb{N}}$ be a sequence of solutions of (3.35) for corresponding $(\varepsilon_n)_{n \in \mathbb{N}}$ with $\varepsilon_n \searrow 0$. Then we deduce*

$$\|z_n\|_\infty \leq C \tag{3.37}$$

with a constant C independent of $n \in \mathbb{N}$.

Proof. Let $(\varepsilon_n, z_n)_{n \in \mathbb{N}}$ denote a sequence of solutions of (3.35) and we assume for any constant $K > 0$ the existence of a certain $x \in [0,1]$ and $n \in \mathbb{N}$ such that $\|z_n(x)\| > K$. In particular we obtain $x_m \in [0,1]$ and n_m such that $\|z_{n_m}(x_m)\| > m$ for every $m \in \mathbb{N}$. With respect to the continuity of z_{n_m} we know, that

$$x \mapsto \|z_{n_m}(x)\| \text{ has a global maximum in } [0,1]$$

and without loss of generality we assume this maximum at $x = x_m$. After selecting a subsequence if necessary (not relabelled), we assume furthermore without loss of generality $\|z_{n_m}(x_m)\| = z_{n_m}(x_m) > m$ and in particular we have

$$\lim_{m \to \infty} z_{n_m}(x_m) = \infty. \tag{3.38}$$

Due to the shape of σ, (3.38) implies

$$\lim_{m \to \infty} \sigma(z_{n_m}(x_m)) = \infty. \tag{3.39}$$

Moreover we have for every $m \in \mathbb{N}$ (after a continuous continuation at $\{0,1\}$ if necessary)

$$-\varepsilon_{n_m} z_{n_m}''(x_m) + \sigma(z_{n_m}(x_m)) = \int_{x_m}^1 b(\lambda, u_{n_m}(\tau), \tau) d\tau. \tag{3.40}$$

The growth conditions on b together with (3.36) imply for every $m \in \mathbb{N}$ due to the continuous embedding $W^{1,p+1}(0,1) \hookrightarrow C^0(0,1)$

$$\left\| \int_{x_m}^1 b(\lambda, u_{n_m}(\tau), \tau) d\tau \right\| \leq \int_{x_m}^1 \|b(\lambda, u_{n_m}(\tau), \tau)\| \, d\tau \leq$$
$$\leq c_3 \|u_{n_m}\|_\infty^r + c_4 \leq c_3 \|u_{n_m}\|_{1,p+1}^r + c_4 \leq c_5.$$

Therefore we have by (3.38) and (3.39)

$$\varepsilon_{n_m} z_{n_m}''(x_m) \geq \sigma(z_{n_m}(x_m)) - c_5 \to \infty$$

for $m \to \infty$ and in particular we get due to $\varepsilon_{n_m} \searrow 0$ the estimate $z_{n_m}''(x_m) > 0$ for m sufficiently large. On the other hand we have $z_{n_m}''(x_m) = 0$ which is obviously true because x_m is assumed to be a global maximum of z_{n_m} and $z_{n_m}'(0) = z_{n_m}'(1) = 0$ is automatically fulfilled. Thus we obtain, that x_m cannot be a maximum contradicting our assumption. $\square$

We will prove some simple consequences of Lemma 22:

Lemma 23. *Let $(z_n)_{n \in \mathbb{N}}$ be a sequence of solutions of (3.35) for corresponding $(\varepsilon_n)_{n \in \mathbb{N}}$ as $\varepsilon_n \searrow 0$. Then we obtain the following:*

(i) The sequence $(\varepsilon_n z_n'')_{n \in \mathbb{N}}$ is uniformly bounded in $C^0([0, 1])$.

(ii) The sequence $(\sqrt{\varepsilon_n} z_n')_{n \in \mathbb{N}}$ is uniformly bounded in $L^2(0, 1)$.

(iii) We have $\varepsilon_n z_n'' \rightharpoonup^* 0$ in $L^\infty(0, 1)$ and

(iv) $\varepsilon_n z_n' \to 0$ in $C^0([0, 1])$.

Proof. Let $(z_n)_{n \in \mathbb{N}}$ be a sequence with the above mentioned property. Further note, that (i) is trivial by (3.35) and Lemma 22. To prove (ii), we multiply (3.35) by z_n, integrate the equation over the unit interval and integration by parts yields

$$\int_0^1 \varepsilon_n z_n'^2 dx = \int_0^1 \left(\int_x^1 b(\lambda, u_n(\tau), \tau) d\tau z_n - \sigma(z_n) z_n \right) dx,$$

which proves immediately the assertion.
The third item is established as follows: Let $\varphi \in C_0^\infty(0, 1)$, then we get again by integration by parts and (ii)

$$\left\| \int_0^1 \varepsilon_n z_n'' \varphi dx \right\| \leq \sqrt{\varepsilon_n} \int_0^1 \|\sqrt{\varepsilon_n} z_n' \varphi'\| \, dx \leq \sqrt{\varepsilon_n} \|\sqrt{\varepsilon_n} z_n'\|_{L^2} \|\varphi'\|_{L^2} \to 0$$

for $n \to \infty$. Because φ was chosen arbitrary the claimed result follows.
Due to (3.35) and the boundary conditions we obtain by virtue of (iii)

$$\|\varepsilon_n z_n'(x)\| = \left\| \int_0^x \varepsilon_n z_n''(s) ds \right\| \to 0. \tag{3.41}$$

Furthermore we have $(\varepsilon_n z_n)_{n \in \mathbb{N}}$ as well as $(\varepsilon_n z_n'')_{n \in \mathbb{N}}$ is bounded in $C^0(0, 1)$ and standard interpolation yields $(\varepsilon_n z_n)_{n \in \mathbb{N}}$ is bounded in $C^2(0, 1)$. Compact imbedding $C^2([0, 1]) \hookrightarrow C^1([0, 1])$ together with (3.41) implies (iv). $\quad\square$

3.5 Geometric structure of $\mathcal{C}^+$

In order to obtain better bounds for $(z_n)_{n \in \mathbb{N}}$ to establish convergence in some function–space we consider in the sequel of the chapter a body force potential b with the following properties:
There exists $x_0 \in (0, 1)$ depending only on λ, such that for every $\lambda \in \mathbb{R}^+$ and $u \in \mathbb{R}$ we have

(H-1) $b(\lambda, u, \cdot)_{|[0, x_0]} \geq 0$.

(H-2) $b(\lambda, u, \cdot)_{|[x_0, 1]} \leq 0$.

(H-3) Furthermore there exists open subsets $\Omega_1 \subseteq [0, x_0]$ and $\Omega_2 \subseteq (x_0, 1]$ with $|\Omega_j| > 0$ for $j = 1, 2$ and $b(\lambda, u, \cdot)_{|\Omega_1} > 0$ and $b(\lambda, u, \cdot)_{|\Omega_2} < 0$.

(H-4) There exists $\vartheta > 0$ such that $b(\lambda, u, \cdot)_{|(0, \vartheta)} > 0$ for every $\lambda \in \mathbb{R}^+$ and $u \in \mathbb{R}$.

Remark 24. We can especially allow a "live" body–force b which has degenerate zeros. Moreover $b(\lambda, u(0), 0) = 0$ or $b(\lambda, u(1), 1) = 0$ is also allowed.

Example 25. The above assumptions are valid for the following body–force: Let W be some "one–well" potential satisfying the growth conditions with $p = 3$ and let $b(\lambda, u, x) := c(\lambda, x)(1 + u^2)$ with $c \in C(\mathbb{R} \times [0,1], \mathbb{R})$ and c changes sign from $+$ to $-$ at $x_0 = x_0(\lambda)$ for $\lambda > 0$.

Let $\mathcal{C}_\varepsilon^+$ be the unbounded path obtained by the global implicit function theorem for fixed $\varepsilon > 0$ and $\lambda > 0$ and let $z \in \mathcal{C}_\varepsilon^+$. Our goal is to show that z' changes sign in $(0,1)$ at most once. For the expression "change sign" we propose the same definition as in [2]:

Definition 26. A function v has exactly k changes of sign in $[0,1]$ provided:

(i) There are points $s_1 < s_2 < \cdots < s_{k+1} \in [0,1]$ with $v(s_i)v(s_{i+1}) < 0$ for all $i = 1, \dots, k$.

(ii) We will say that v changes sign at the points $x_1 < \cdots < x_k$ provided $v(x_i) = 0$ and the points in (i) can be chosen so that $s_i < x_i < s_{i+1}$ and v does not change sign in $[x_i, x_{i+1}]$.

Remark 27. Before continuing our work let us fix some notation for our convenience: We fix $\varepsilon > 0$ and let $\hat{\lambda} \in \mathbb{R}^+$, then $z = z(\hat{\lambda}) = z_{\hat{\lambda}}$ is the solution on the branch of

$$u' = z,$$
$$-\varepsilon z'' + \sigma(z) = \int_x^1 b(\hat{\lambda}, u(\tau), \tau)d\tau,$$
$$u(0) = z'(0) = z'(1) = 0.$$

Let us further remark that for our forthcoming analysis differentiation of (3.35) turns out to be very useful. For the reader's convenience we denote the differentiated form of (3.35):

$$h = z',$$
$$\varepsilon h'' - \sigma'(z)h = b(\lambda, u, \cdot), \tag{3.42}$$
$$h(0) = h(1) = 0.$$

For λ_0 sufficiently small, it is easy to see that $\sigma'(z_\lambda) > 0$ for every $0 \le \lambda \le \lambda_0$ because of our assumptions on the stored–energy density W and $z_{|\{\lambda=0\}} \equiv 0$. Hence we derive the following lemma:

Lemma 28. Let $z_\lambda \in \mathcal{C}_\varepsilon^+$ and consider λ_0 such that $\sigma'(z_\lambda)_{|\lambda \in [0,\lambda_0]} > 0$. Then z'_λ changes sign at most once for $0 < \lambda \le \lambda_0$.

Proof. We prove by contradiction and of course it is enough to assume, that z'_λ changes sign exactly two times in $[0,1]$. By Definition 26 there exists points $s_1, s_2, s_3 \in (0,1)$ with $z'_\lambda(s_i)z'_\lambda(s_{i+1}) < 0$ and $x_1, x_2 \in (0,1)$ such that $s_1 < x_1 < s_2 < x_2 < s_3$ and $z'_\lambda(x_1) = z'_\lambda(x_2) = 0$.
With respect to the boundary condition $z'_\lambda(0) = z'_\lambda(1) = 0$ we have three nodal areas of z'_λ in $[0,1]$. On the other hand we have just one sign change of b and hence, it is obvious that there exists a nodal area $\Omega' \subseteq [0,1]$ of z'_λ such that

$$\text{sign } b_{|\Omega'} = \text{sign } z'_{\lambda|\Omega'}. \tag{3.43}$$

For every $x \in \Omega'$ we have by (3.42)

$$\varepsilon(z_\lambda')'' - \sigma'(z_\lambda)z_\lambda' = b, \\ z_{\lambda|\partial\Omega'}' = 0. \tag{3.44}$$

By the weak maximum principle we conclude

$$\operatorname{sign} b_{|\Omega'} \neq \operatorname{sign} z_{\lambda|\Omega'}'$$

contradicting (3.43). $\qquad\square$

By virtue of Lemma 28 there are two possibilities for the behavior of z_λ' for $0 < \lambda < \lambda_0$:

(i) z_λ' has exactly one sign change at $x_1 \in (0,1)$.

(ii) There is no sign change of z_λ'.

For (i) we have to distinguish between the cases

(a) $x_1 \in (x_0, 1)$,

(b) $x_1 \in (0, x_0]$.

Also for (ii) we obtain two possibilities:

(a) $z_\lambda'(x) \leq 0$ or

(b) $z_\lambda'(x) \geq 0$ for every $x \in [0,1]$.

First we state a property of an arbitrary solution of the Euler–Lagrange equation without using any continuation method. Nevertheless this observation will be very useful for the method.

Lemma 29. *Let z be a solution of* (3.35). *Then there is no negative minimum of z' in* $(x_0, 1)$.

Proof. Suppose not. Then we get existence of $x_1 \in (x_0, 1)$ with

$$z'(x_1) < 0,\ z''(x_1) = 0,\ z'''(x_1) \geq 0.$$

With respect to the identity

$$-\sigma'(z(x_1))z'(x_1) = -\varepsilon z'''(x_1) + b(\lambda, u(x_1), x_1) \leq 0$$

we obtain

$$z(x_1) \in [\nu_1, \nu_2] \subseteq \mathbb{R}^+. \tag{3.45}$$

On the other hand we have the identity

$$\sigma(z(x_1)) = \int_{x_1}^1 b(\lambda, u(\tau), \tau)d\tau \leq 0,$$

which yields $z(x_1) \leq 0$ and this contradicts (3.45). $\qquad\square$

Before using the continuation method to obtain certain properties of solutions of the Euler–Lagrange equation for arbitrary $\lambda > 0$ we give another result for z'_λ, λ sufficiently small.

Lemma 30. *Let λ_0 be defined as in Lemma 28 and let $z_\lambda \in \mathcal{C}^+_\varepsilon$ for $\lambda < \lambda_0$ and an arbitrary $\varepsilon > 0$. Then there exists no interval $J \subseteq [0, x_0]$ such that $z'_{\lambda | J} \equiv 0$.*

Proof. Suppose not. Then we obtain an interval $J := [x_\alpha, x_\beta] \subseteq [0, x_0]$ with $z'_{\lambda | J} \equiv 0$. For the sequel of the proof we omit λ for our convenience. Let $\hat{x} \in (x_\alpha, x_\beta)$ and $z'_{|(0,\hat{x})}$ satisfies the equation

$$\varepsilon(z')'' - \sigma'(z)z' = b \geq 0$$
$$z'(0) = z'(\hat{x}) = z''(\hat{x}) = 0. \tag{3.46}$$

The weak maximum principle tells us $z'_{|(0,\hat{x})} \leq 0$ and the maximum principle due to Hopf considering the derivative on the boundary implies $z'_{|(0,\hat{x})} \equiv 0$ and in particular $z'_{|(0,x_\beta)} \equiv 0$. This implies immediately by (3.46) $b_{|(0,x_\beta)} \equiv 0$ which is impossible by our assumptions on b (in particular (H–4)). $\qquad\square$

The next lemma tells us necessary conditions for the existence of an interval $J \subseteq (x_0, 1)$ such that $z'_{|J} \equiv 0$.

Lemma 31. *Let $\lambda \in (0, \lambda_0)$ and $z_\lambda \in \mathcal{C}^+_\varepsilon$ for an arbitrary but fixed $\varepsilon > 0$ and we assume the existence of an interval $J \subseteq (x_0, 1)$ such that $z'_{\lambda | J} \equiv 0$. Then we obtain $z'_\lambda(x) \leq 0$ for every $x \in [0, 1]$. Moreover there exists $\bar{x} \in (x_0, 1)$ with $z'_{\lambda | (0,\bar{x})} < 0$ and $z'_{\lambda | (\bar{x},1)} \equiv 0$.*

Proof. Let $J := [x_\alpha, x_\beta] \subseteq (x_0, 1)$ an interval with $z'_{|J} \equiv 0$ (where we omitted again λ). In an analogues manner as in the proof of Lemma 30 one can show that this implies

$$z'_{|[x_\alpha,1]} \equiv 0.$$

Furthermore our assumptions on b exclude the possibility, that $z'_{|(x_0,1)} \equiv 0$ by the same reasons given in the proof of Lemma 30.

Thus there exists an $x \in (x_0, 1)$ such that $z'(x) \neq 0$ and let $\bar{x} \in (x_0, 1)$ such that $z'(\bar{x}) = z''(\bar{x}) = 0$ and $z'(x) \neq 0$ for $x \in (\bar{x} - \delta, \bar{x})$ for some $\delta > 0$. Hopf's maximum principle applied to (3.42) implies immediately

$$z'_{|(\bar{x}-\delta,\bar{x})} < 0.$$

Furthermore the weak maximum principle tells us that there is no $x_\beta \in (x_0, \bar{x})$ such that $z'(x_\beta) = 0$. Hence we conclude

$$z'_{|(x_0,1)} \leq 0, \quad z'_{|(x_0,\bar{x})} < 0.$$

The proof is done if we can prove, that there is no $x \in (0, x_0)$ with $z'(x_0) \geq 0$. We assume this and first we consider $x' < x'' < x_0$ such that $z'(x') = z'(x'') = 0$ and $z'_{|(x',x'')} > 0$. We use the equation

$$\varepsilon(z')'' - \sigma'(z)z' = b \geq 0,$$
$$z'(x') = z'(x'') = 0. \tag{3.47}$$

By the weak maximum principle we obtain $z'_{|(x',x'')} \leq 0$ and thus we have shown

$$z'_{|(0,\bar{x})} \leq 0.$$

To finish the proof we just have to show that there is no $x \in (0,x_0)$ such that $z'(x) = z''(x) = 0$. But this is again a consequence of Hopf's maximum principle used for equation (3.47) and applied to the interval $(0,x)$. □

With Lemma 30 and 31 at hand we can easily refine our possibilities for z' stated after Lemma 28.

Corollary 32. *Choose an arbitrary but fixed $\varepsilon > 0$. Furthermore consider $z_\lambda \in \mathcal{C}^+_\varepsilon$ and λ_0 chosen so that $\sigma'(z_\lambda) > 0$ for $\lambda \in (0,\lambda_0)$ is valid. Then we have the following two possibilities:*

(i) z'_λ has exactly one sign change and moreover we know $z''_\lambda(0) < 0$, $z''_\lambda(1) < 0$ and the only zero at $x = x_1$ of z'_λ is nondegenerate with $z''_\lambda(x_1) > 0$. In particular, the zero of z'_λ in $(0,1)$ is unique.

(ii) z'_λ does not change sign and if there exists an interval $J \subseteq (0,1)$ such that $z'_{\lambda|J} \equiv 0$, then $J = (\bar{x},1]$ with $\bar{x} > x_0$ and $z'_{\lambda|(0,\bar{x})} < 0$.

Proof. Let λ_0 be defined as above and we omit λ in the sequel of the proof.

(i) We assume, that z' has exactly one sign change at x_1. By virtue of Lemma 30 and 31 there exists no interval J with $z'_{|J} \equiv 0$ and the weak maximum principle again applied to (3.42) implies immediately $z'_{|(0,x_1)} \leq 0$ and $z'_{|(x_1,1)} \geq 0$.
Without loss of generality we assume $(0,x_1) \subseteq (0,x_0]$. By considering again (3.47) with corresponding boundary conditions $z'(0) = 0$ and $z'(x_1) = 0$, Hopf's maximum principle yields

$$z'_{|(0,x_1)} < 0, \ z''(0) < 0 \text{ and } z''(x_1) > 0.$$

Again by Hopf's maximum principle we deduce $z'_{|(x_0,1)} > 0$ and $z''(1) < 0$. Moreover we have $z'_{|(x_1,x_0)} > 0$ because of the weak maximum principle.

(ii) This case is identical to Lemma 30 and 31.

□

In the sequel of this section we propose a continuation method to obtain our above mentioned goal: The function z'_λ has at most one sign change in $(0,1)$ for every $\lambda \in \mathbb{R}^+$. Because of $\mathcal{C}^+_\varepsilon$ is connected in the C^2–topology we deduce, that z'_λ is connected in the C^1–topology. Moreover we have the following:

Proposition 33. *There are just two (respectively three) possibilities how to create a new zero on a C^1–component:*

1. *There exists $x_2 \in (0,1)$ such that $z'(x_2) = z''(x_2) = 0$ and moreover we have $\text{sign}(z'(x_2 - \gamma)) = \text{sign}(z'(x_2 + \gamma))$ for every $\gamma \in (0,\kappa)$, κ sufficiently small.*

2. *To create a new zero at the boundary we assume either $z''(0) = 0$ or $z''(1) = 0$.*

3. *Furthermore in case (i) of Corollary 32 the following situation can occur: Let z' change sign at $x = x_1$. Then it is possible to have $z'(x_1) = z''(x_1) = 0$.*

In the forthcoming analysis we will often use the well–known generalized maximum principle and we will state the theorem for the reader's convenience. For the proof we refer to [36].

Theorem 34. *Let Ω be an open domain in $\mathbb{R}^n$ with smooth boundary and let $w \in C^2(\Omega)$ be a function satisfying*

$$\triangle w + c(x)w \geq 0,$$

where the sign of $c(x)$ does not matter. Further on we assume $w \leq 0$ and $\max\limits_{x \in \Omega} w(x) = w(x_0) = 0$. Moreover we assume one of the two possibilities:

(i) $x_0 \in \Omega$,

(ii) $x_0 \in \partial\Omega$ and $\frac{\partial}{\partial n} w(x_0) = 0$.

Both cases imply $w \equiv 0$.

Now we can derive the main statement of the section:

Theorem 35. *Let $\varepsilon > 0$ arbitrarily but fixed and consider an arbitrary $z \in C_\varepsilon^+$. Then we obtain the following three possibilities for the shape of z' and obviously they exclude each other:*

(i) The function z' has exactly one change in sign at $x = x_1$ and moreover we have $z'_{|(0,x_1)} < 0$ and $z'_{|(x_1,1)} > 0$. Furthermore $z''(0) < 0$, $z''(x_1) > 0$ and $z''(1) < 0$ holds.

(ii) We have $z'(x) \leq 0$ for every $x \in [0,1]$ and if there exists an interval $J \subseteq [0,1]$ such that $z'_{|J} \equiv 0$, then $J = [x_\alpha, 1]$ for a certain $x_\alpha > x_0$.

(iii) For every $x \in (0,1)$ we obtain $z'(x) > 0$.

Proof. We seperate the proof into two steps:

Step 1:   First we consider case (i) of Corollary 32: Let x_1 be the nondegenerate zero of z'_{λ_1} for some λ_1 sufficiently small. By virtue of Corollary 32 we have

$$z'_{\lambda_1|(0,x_1)} < 0 \text{ and } z'_{\lambda_1|(x_1,1)} > 0.$$

Our analysis forces us to distinguish (i) in two further cases:

(a) Either $x_1 \in [x_0, 1)$

(b) or $x_1 \in [0, x_0)$ holds.

First we establish our analysis for (a):

Suppose the existence of $\lambda_2 > \lambda_1$ and $x_2 \in (0, x_0)$ with $z'_{\lambda_2}(x_2) = z''_{\lambda_2}(x_2) = 0$ and moreover we have $\mathrm{sign}(z'_{\lambda_2}(x_2 - \gamma)) = \mathrm{sign}(z'_{\lambda_2}(x_2 + \gamma))$ for every $\gamma \in (0, \kappa)$, κ sufficiently small. We know by Corollary 32 and the continuation method $z'_{\lambda_2|(0,x_2)} < 0$ and furthermore we have

$$\varepsilon(z')'' - \sigma'(z)z' = b,$$
$$z'(0) = z'(x_2) = 0.$$

Theorem 34 yields $z'_{\lambda_2|[0,x_2]} \equiv 0$, which is impossible by the assumption (H–4) on b.

If we assume the existence of $x_3 \in (x_0, x_1)$ with $z'_{\lambda_2}(x_3) = z''_{\lambda_2}(x_3) = 0$, then we will have two possibilities:

The first one is the following: There exists $x' \in (x_3, x_1)$ such that z'_{λ_2} possesses a negative minimum at x', which is shown in the following picture:

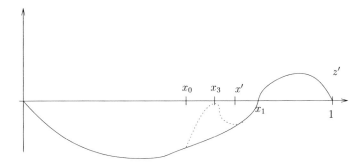

But this situation is impossible by Lemma 29. The other possibility is that $z'_{\lambda_2|(x_3,1)} \equiv 0$ and we obtain $z'_{\lambda_2} \leq 0$, a case which will be discussed in step 2 of the proof. Furthermore we obtain $z'_{\lambda_2|(x_4,1)} \equiv 0$ and $z'_{\lambda_2} \leq 0$ by virtue of Theorem 34 if we consider $x_4 \in (x_1, 1)$ with $z'_{\lambda_2}(x_4) = z''_{\lambda_2}(x_4) = 0$.

Now we consider the creation of a new zero at the boundary: First we observe that $z''_{\lambda_2}(0) = 0$ for some $\lambda_2 > \lambda_1$ is impossible because under this assumption Theorem 34 applied to (3.42) yields $z'_{\lambda_2|(0,x_0)} \equiv 0$. Again by virtue of Theorem 34 $z''_{\lambda_2}(1) = 0$ predicts $z'_{\lambda_2} \leq 0$.

If we consider the case $z'_{\lambda}(x_1) = z''_{\lambda}(x_1) = 0$ for some $\lambda > 0$, where x_1 denotes the unique zero of z'_{λ} in $(0, 1)$, we have the following: By assumption $x_1 \in (x_0, 1)$ and therefore we obtain due to the fact that $z'_{\lambda|(x_1,1)} \geq 0$ and Theorem 34

$$z'_{\lambda|(x_1,1)} = 0$$

and again we have $z'_{\lambda} \leq 0$ which will be treated in step 2 of the proof.

We consider (b), which means existence of $\lambda_1 \in (0, \lambda_0)$ and $x_1 \in (0, x_0]$ such that $z'_{\lambda_1|(0,x_1)} < 0$ and $z'_{\lambda_1|(x_1,1)} > 0$:

First we exclude possibility one of Proposition 33: If there exists $x_2 \in (0, x_1)$ respectively $x_3 \in (x_0, 1)$ with $z'_{\lambda_2}(x_j) = z''_{\lambda_2}(x_j) = 0$ for $j = 2, 3$ and for some $\lambda_2 > \lambda_1$, we obtain by Theorem 34 and

$$\varepsilon(z'_{\lambda_2})'' - \sigma'(z_{\lambda_2})z'_{\lambda_2} = b,$$
$$z'_{\lambda_2}(0) = z'_{\lambda_2}(x_j) = z''_{\lambda_2}(x_j) = z'(1) = 0$$

for $j = 2, 3$ the following:

$$z'_{\lambda_2|(0,x_1)} \equiv 0 \text{ respectively } z'_{\lambda_2|(x_0,1)} \equiv 0.$$

Both is impossible by virtue of (3.42) and the assumption (H–3) on b.

The case $x_4 \in (x_1, x_0)$ with $z'_{\lambda_2}(x_4) = z''_{\lambda_2}(x_4) = 0$ for some $\lambda_2 > \lambda_1$ involves a bit more difficulties because a maximum principle does not apply and we show the typical situation in the picture:

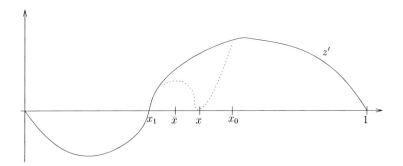

First we observe by the continuation method the existence of $\bar{x} \in (x_1, x_4)$ with

$$z'_{\lambda_2}(\bar{x}) > 0, \quad z''_{\lambda_2}(\bar{x}) = 0, \quad z'''_{\lambda_2}(\bar{x}) \leq 0.$$

By the identity

$$-\sigma'(z_{\lambda_2}(\bar{x}))z'_{\lambda_2}(\bar{x}) = b(\lambda_2, u(\bar{x}), \bar{x}) - \varepsilon z'''_{\lambda_2}(\bar{x}) \geq 0$$

we deduce $\sigma'(z_{\lambda_2}(\bar{x})) \leq 0$ and therefore we obtain

$$z_{\lambda_2}(\bar{x}) \in [\nu_1, \nu_2] \subseteq \mathbb{R}^+.$$

Because of $z'_{\lambda_2|(x_1,1)} \geq 0$ we know $z_{\lambda_2}(1) \geq z_{\lambda_2}(\bar{x}) \geq \nu_1 > 0$ and by virtue of

$$\sigma(z_{\lambda_2}(1)) = \varepsilon z''_{\lambda_2}(1)$$

we get $z''_{\lambda_2}(1) > 0$ which yields

$$z'_{\lambda_2|(1-\varrho,1)} < 0 \quad \text{for } \varrho > 0 \text{ sufficiently small.}$$

By the first part of the proof of (b) it is impossible that z' changes sign in $(x_0, 1)$ and moreover we know due to our continuation method $z'_{\lambda_2|(x_0,1)} > 0$, hence a contradiction.

Next we exclude the creation of new zeros at the boundary: We assume existence of $\lambda_3 > \lambda_1$ with $z''_{\lambda_3}(0) = 0$ respectively $z''_{\lambda_3}(1) = 0$. Both cases imply by Theorem 34 applied to equation (3.42) $z'_{\lambda_3 | (0, x_1)} \equiv 0$ respectively $z'_{\lambda_3 | (x_0, 1)} \equiv 0$ which is impossible due to the assumptions on b.

Excluding possibility three of Proposition 33 is trivial: We consider $z'_{\lambda_4}(x_1) = z''_{\lambda_4}(x_1) = 0$ for some $\lambda_4 > \lambda_1$ and we obtain again by Theorem 34 $z'_{|(0, x_1)} \equiv 0$, impossible by the assumption (H–4) on b.

Hence we derived the following:

Proposition 36. *Let $\lambda_1 \in \mathbb{R}$ such that z'_{λ_1} changes sign exactly once with nondegenerate zeros. Then we have two possibilities:*

1. *z'_λ has the same properties as z'_{λ_1} for every $\lambda > \lambda_1$.*

2. *We obtain existence of $\lambda_2 > \lambda_1$ with $z'_{\lambda_2} \leq 0$. Moreover we have $z''_{\lambda_2}(0) < 0$ and if there exists an interval $J \subseteq [0,1]$ with $z'_{\lambda_2 | J} \equiv 0$, then $J = [x_2, 1]$ for some $x_2 > x_0$.*

Step 2: We consider the second case of Corollary 32:

First we treat $z'_{\lambda_1}(x) \leq 0$ for a certain $\lambda_1 \in \mathbb{R}^+$ and for every $x \in [0,1]$. (Note, that it is not necessary to assume λ_1 to be sufficiently small). We have to show that z'_λ changes sign at most once for arbitrary $\lambda > \lambda_1$.

Again by Theorem 34 applying to (3.42) we exclude existence of a solution $z_\lambda \in \mathcal{C}^+_\varepsilon$ satisfying $z''_\lambda(0) = 0$. By the same reason existence of $x_1 \in (0, x_0]$ with $z'_\lambda(x_1) = z''_\lambda(x_1) = 0$ is impossible.

Let $x_\alpha \in (x_0, 1]$ defined in such a way that

$$z'_{\lambda_1 | (0, x_\alpha)} < 0 \text{ and } z'_{\lambda_1 | (x_\alpha, 1]} \equiv 0$$

is valid. Furthermore we consider $x_2 \in (x_0, x_\alpha)$ with

$$z'_{\lambda_2}(x_2) = z''_{\lambda_2}(x_2) = 0 \text{ for some } \lambda_2 > \lambda_1.$$

If $z'_{\lambda_2 | (x_2, 1)} \neq 0$ there has to be a negative minimum of z'_{λ_2} in $(x_0, 1)$ which contradicts Lemma 29 and therefore we obtain

$$z'_{\lambda_2 | (x_2, 1)} \equiv 0.$$

After these observations we conclude by (3.42) that no sign change of z'_λ will occur in the interval $(0, x_\beta)$ with

$$x_\beta(\lambda) := \inf \{\hat{x} \mid b(\lambda, u_\lambda(x), x) \equiv 0 \quad \text{for every } x \geq \hat{x}\} > x_0,$$

where $u_\lambda = z'_\lambda$.

Let $\hat{\lambda} > \lambda_1$ and $z'_{\hat{\lambda} | (x_\gamma, 1]} \equiv 0$ for some $x_\gamma(\hat{\lambda}) \geq x_\beta(\hat{\lambda})$. We have to exclude the occurrence of oscillation of z'_λ in $(x_\beta(\lambda), 1]$ for $\lambda > \hat{\lambda}$, which is easily done by Lemma 29. Note also, that $z'_{\lambda | (x_\beta(\lambda), 1]}$ can change sign at most once according to Lemma 29. Moreover, if a change in sign takes place at $x = x_1$, then we obtain by Theorem 34 applied to (3.42) $z'_{\lambda | (x_1, 1)} > 0$, $z'_{\lambda | (0, x_1)} < 0$ and every zero of z'_λ is non–degenerate.

Therefore we proved the following:

Proposition 37. *Let $\lambda_1 \in \mathbb{R}$ such that $z'_{\lambda_1} \leq 0$. Then we have two possibilities:*

1. *For every $\lambda > \lambda_1$ any solution $z_\lambda \in \mathcal{C}^+_\varepsilon$ satisfies the property $z'_\lambda \leq 0$.*

2. *There exists $\lambda_2 > \lambda_1$ such that z'_{λ_2} changes sign exactly once at $x = x_1$ and we have $z'_{\lambda_2|(0,x_1)} < 0$ and $z'_{\lambda_2|(x_1,1)} > 0$. Moreover all zeros of z'_{λ_2} are non degenerate.*

By virtue of Proposition 36 and 37 we obtain immediately

Proposition 38. *Let $\lambda_1 \in \mathbb{R}^+$ such that either z'_{λ_1} changes sign exactly once from $-$ to $+$, where the zeros are non degenerate or $z'_{\lambda_1} \leq 0$. Furthermore let $z_\lambda \in \mathcal{C}^+_\varepsilon$ for an arbitrary $\lambda > \lambda_1$. Then we have one of the following possibilities:*

1. *The function z'_λ changes sign at $x = x_1$ and every zero of z'_λ is non–degenerate.*

2. *For every $x \in [0,1]$ we obtain $z'_\lambda(x) \leq 0$.*

It remains to consider the case $z'_{\lambda_1|(0,1)} \geq 0$ for a certain $\lambda_1 > 0$ and we have to show that z'_λ changes sign at most once for every $\lambda > \lambda_1$.

We assume for the sequel of the proof the existence of $\lambda_2 > \lambda_1$ such that z'_{λ_2} changes sign at least twice: Then there exists a local positive maximum at $x = x_\alpha$ and a local negative minimum at $x = x_\beta$:

Note, that $x_\alpha \leq x_0$ holds, because if we assume the opposite we get due to the number of zeros of z'_{λ_2} the existence of a local negative minimum of z'_{λ_2} in $[x_0, 1)$ contradicting Lemma 29. By virtue of $z'_{\lambda_2}(x_\alpha) > 0$ and $z'''_{\lambda_2}(x_\alpha) \leq 0$ we obtain

$$-\sigma'(z_{\lambda_2}(x_\alpha))z'_{\lambda_2}(x_\alpha) \geq \varepsilon z'''(x_\alpha) - \sigma'(z_{\lambda_2}(x_\alpha))z'_{\lambda_2}(x_\alpha) = b(\lambda_2, u(x_\alpha), x_\alpha) \geq 0.$$

In particular this implies $z_{\lambda_2}(x_\alpha) \in [\nu_1, \nu_2] \subseteq \mathbb{R}^+$. Thus a necessary condition for the existence of a local positive maximum and a local negative minimum of z'_{λ_2} is the existence of $x \in [0,1]$ such that

$$z_{\lambda_2}(x) \geq \nu_1.$$

On the other hand we have by Theorem 34

$$0 > \varepsilon z''_{\lambda_1}(1) = \sigma(z_{\lambda_1}(1)) \tag{3.48}$$

and this implies due to the assumed monotonicity

$$z_{\lambda_1}(x) < 0 \text{ for every } x \in [0,1].$$

Of course, z_λ is connected with respect to the C^1–topology and we obtain by virtue of Proposition 38 the existence of $\hat{\lambda} \in (\lambda_1, \lambda_2)$ such that $z_{\hat{\lambda}}$ has the following properties:

$$z'_{\hat{\lambda}|(0,1)} \geq 0 \text{ and there exists } \hat{x} \in [0,1] \text{ with } z_{\hat{\lambda}}(\hat{x}) = 0. \tag{3.49}$$

We consider again the equation

$$\begin{aligned} \varepsilon(z')'' - \sigma'(z)z' &= b, \\ z'(0) = z'(1) &= 0 \end{aligned} \tag{3.50}$$

for $z = z_{\hat{\lambda}}$.

Due to $z'_\lambda \geq 0$ we obtain $z''_\lambda(1) < 0$, because otherwise Theorem 34 and (3.50) would imply

$$z'_{\lambda_{|(x_0,1)}} \equiv 0$$

in contradiction to our assumptions on b. As in (3.48) we conclude $z_{\hat\lambda}(1) < 0$. Again by the assumed monotonicity we get

$$z_{\hat\lambda}(x) < 0 \quad \text{for every } x \in [0,1],$$

thus a contradiction to (3.49).
Hence we proved the following:

Proposition 39. *Let $\lambda_1 \in \mathbb{R}$ such that $z'_{\lambda_1} \geq 0$. Then we have two possibilities:*

1. *We have for every $\lambda > \lambda_1$ the property $z'_\lambda \geq 0$.*

2. *There exists $\lambda_2 > \lambda_1$ such that z'_{λ_2} changes sign exactly once. Note, that the zeros of z'_{λ_2} are non degenerate by the maximum principle.*

Thus, by virtue of Proposition 36, 37 and 39 we obtain, that z'_λ changes sign at most once and moreover (i) and (ii) of Theorem 35 is valid.

It remains to show the third item of Theorem 35 and therefore we assume

$$z'(x) \geq 0 \quad \text{for every } x \in [0,1]$$

and moreover existence of $\hat{x} \in (0,1)$ such that $z'(\hat{x}) = 0$. Theorem 34 applied to (3.42) yields, that we have no degenerate zero of z' in $[x_0, 1)$. If we assume $\hat{x} \in (0, x_0)$ we obtain the following: First note, that the assumptions on b and (3.42) imply

$$z'_{|(0,\delta)} \text{ is not identically zero.}$$

for some $\delta > 0$. Hence by our assumption on z' we get existence of a local maximum $x_2 \in (0, x_1)$ with $z'(x_2) > 0$. By the same procedure as in the proof of Proposition 39 we obtain a contradiction. Hence $z'(x) > 0$ for every $x \in (0,1)$, which completes the proof of Theorem 35. □

3.6 Singular limit analysis

Let $z_{\varepsilon_n} \in \mathcal{C}^+_{\varepsilon_n}$ for every $n \in \mathbb{N}$. Our goal here is to pass to the limit for the sequence $(z_{\varepsilon_n})_{n \in \mathbb{N}}$ as ε_n tends to zero. This is essentially done by Theorem 35 and Hellys theorem. Afterwards we show some qualitative properties of this limit.

Lemma 40. *Consider a sequence $(z_n)_{n \in \mathbb{N}} \in \mathcal{C}^+_{\varepsilon_n}$ for fixed $\lambda \in \mathbb{R}^+$ and $\varepsilon_n \searrow 0$. We obtain*

$$TV(z_n) \leq C$$

for every $n \in \mathbb{N}$, where TV denotes the total variation and is defined for a function $f : [a,b] \mapsto \mathbb{R}$ by

$$TV(f) := \sup\left\{ \sum_{k=1}^n |f(x_k) - f(x_{k-1})| : a = x_0 < x_1 < \cdots < x_n = b, \, n \in \mathbb{N} \right\}. \quad (3.51)$$

Proof. By Theorem 35 the function z'_n has at most one sign change, say at $x = x(n)$. Due to the fundamental theorem in calculus we obtain by Lemma 22 with a constant K independent of ε

$$||z'_n||_{L^1} = \int_0^{x(n)} -z'_n(s)ds + \int_{x(n)}^1 z'_n(s)ds \leq 4\,||z_n||_{C^0} \leq K.$$

The same bound holds obviously for the total variation by (3.51). □

By Helly's theorem (see for example [34]) we conclude:

Theorem 41. *We consider a sequence $(z_n)_{n\in\mathbb{N}} \in \mathcal{C}^+_{\varepsilon_n}$ with $\varepsilon_n \searrow 0$ for fixed $\lambda \in \mathbb{R}^+$. Then there exists a subsequence (not relabelled) and a function $z \in L^\infty(0,1) \cap BV(0,1)$, such that*

$$\lim_{n\to\infty} z_n(x) = z(x) \tag{3.52}$$

for every $x \in [0,1]$. Here $BV(0,1)$ denotes the space of bounded variation defined on $(0,1)$, see [5] for details.
Moreover we have for every $x \in [0,1]$ by (3.35) and Lemma 23

(i)

$$\lim_{n\to\infty} \varepsilon_n z''_n(x) = 0, \tag{3.53}$$

(ii)

$$\lim_{n\to\infty} \sigma(z_n(x)) = \sigma(z(x)) = \int_x^1 b(\lambda, u(\tau), \tau)d\tau, \tag{3.54}$$

where $u = \int_0^x z(s)ds$ is the limit of $(u_n)_{n\in\mathbb{N}}$ in $C^0(0,1)$. In particular, the first Weierstrass–Erdmann corner condition is fulfilled.

In the sequel of the section we prove some qualitative properties of the function z obtained by Theorem 41.

By the Euler–Lagrange equation we deduce, that z also satisfies the second Weierstrass–Erdmann corner condition. The proof is done in a similar way as in [18].

Corollary 42. *Let z be the limit obtained by Theorem 41. Then the function*

$$x \mapsto W(z(x)) - \sigma(z(x))z(x)$$

is continuous for every $x \in [0,1]$.

Proof. Let $(z_n)_{n\in\mathbb{N}}$ be a sequence like the one considered in Theorem 41 and thus z_n satisfies the equation

$$\varepsilon z'''_n - \sigma'(z_n)z'_n = b.$$

Multiplying this equation by z_n yields (by omitting the subscript n)

$$\frac{\partial}{\partial x}(\varepsilon z''z) - \frac{1}{2}\varepsilon\frac{\partial}{\partial x}(z'^2) - \frac{\partial}{\partial x}(\sigma(z)z) + \frac{\partial}{\partial x}W(z) = b(\lambda, u, \cdot)z.$$

By integrating this equation over the interval $(0, x)$ we obtain

$$
\begin{aligned}
W(z_n(x)) - \sigma(z_n(x))z_n(x) = \\
= \frac{1}{2}\varepsilon_n z_n'(x)^2 - \varepsilon_n z_n''(x)z_n(x) + \int_0^x b(\lambda, u_n(\tau), \tau)z_n(\tau)d\tau + c_n
\end{aligned}
\tag{3.55}
$$

with $c_n := \varepsilon_n z_n''(0)z_n(0) - \sigma(z_n(0))z_n(0) + W(z_n(0))$.

By virtue of Theorem 41 we deduce, that the left hand side of (3.55) converges to $W(z) - \sigma(z)z$ and for the right hand side we claim

$$
\begin{aligned}
\lim_{n\to\infty} \left(\frac{1}{2}\varepsilon_n z_n'(x)^2 - \varepsilon_n z_n''(x)z_n(x) + \int_0^x b(\lambda, u_n(\tau), \tau)z_n(\tau)d\tau + c_n \right) = \\
= \int_0^x b(\lambda, u(\tau), \tau)z(\tau)d\tau + c_0
\end{aligned}
\tag{3.56}
$$

with a constant c_0, where u and z is the limit of $(u_n)_{n\in\mathbb{N}}$ and $(z_n)_{n\in\mathbb{N}}$ respectively. Obviously (3.56) proves the assertion.

Note, that due to (3.53) we obtain $\lim_{n\to\infty} \varepsilon_n z_n''(x)z_n(x) = 0$ and because of Lebesgue's dominated convergence theorem we immediately deduce

$$
\lim_{n\to\infty} \int_0^x b(\lambda, u_n(\tau), \tau)z_n(\tau)\, d\tau = \int_0^x b(\lambda, u(\tau), \tau)z(\tau)d\tau.
$$

Due to the fact that the sequence $(c_n)_{n\in\mathbb{N}}$ is bounded we obtain $c_n \to c_0$ (at least for a subsequence). It remains to show

$$
\lim_{n\to\infty} \varepsilon_n z_n'(x)^2 = 0.
\tag{3.57}
$$

It is easy to see that the sequence $(\varepsilon_n z_n'(x)^2)_{n\in\mathbb{N}}$ converges for every $x \in [0, 1]$ because every term in (3.55) converges. Furthermore we obtain by Lemma 23 and Lemma 40

$$
\left\| \varepsilon_n z_n'^2 \right\|_{L^1(0,1)} \leq \left\| \varepsilon_n z_n' \right\|_\infty \left\| z_n' \right\|_{L^1(0,1)} \to 0
$$

for $n \to \infty$. Together with the above mentioned pointwise convergence of the sequence we obtain (3.57). $\qquad\square$

The next goal is to show that z can jump at most once in the unit interval. The first step in this direction is the observation that the last two cases in Theorem 35 cannot occur if ε tends to zero.

Lemma 43. *Let $(\varepsilon_n)_{n\in\mathbb{N}}$ be an arbitrary sequence converging to zero and let $(z_n)_{n\in\mathbb{N}} \in \mathcal{C}_{\varepsilon_n}^+$ for fixed $\lambda > 0$. Then we obtain for $n \in \mathbb{N}$ sufficiently large*

$$
\mathrm{sign}(z_n') \neq \mathrm{const.}
$$

Proof. We argue by contradiction: By assumption there exists a sequence $(z_n)_{n\in\mathbb{N}}$ with $\mathrm{sign}(z_n') \equiv \mathrm{const.}$ for every $n \in \mathbb{N}$ and furthermore we assume without loss of generality $\mathrm{sign}(z_n') \geq 0$. We obtain by Helly's theorem

$$
\lim_{n\to\infty} z_n(x) = z(x)
$$

for every $x \in [0,1]$, z is monotonically increasing and satisfies the equation

$$\sigma(z(x)) = \int_x^1 b(\lambda, u(\tau), \tau)d\tau$$

with $u(x) = \int_0^x z(s)ds$. With respect to the shape of b we know $\int_x^1 b(\lambda, u(\tau), \tau)d\tau_{|[0,x_0]}$ is monotonically decreasing. By the assumptions on σ and the monotonicity of z and b we obtain

$$z_{|[0,x_0]} \in [\nu_1, \nu_2] \subseteq \mathbb{R}^+.$$

On the other hand we have $\sigma(z(x_0)) = \int_{x_0}^1 b(\lambda, u(\tau), \tau)d\tau < 0$ which yields

$$z(x_0) < 0.$$

Contradiction.   □

In the sequel let $\lambda > 0$ be fixed. We consider a sequence $(\varepsilon_n)_{n \in \mathbb{N}}$ with $\varepsilon_n \searrow 0$ and let $(u_n)_{n \in \mathbb{N}} = ((u_\lambda)_n)_{n \in \mathbb{N}}$ be a sequence of functions on the branch $\mathcal{C}_{\varepsilon_n}^+$ with $u_n \to u$ uniformly in $C^0(0,1)$. Because of the assumptions (H-1) and (H-2) on b we define for $\delta > 0$ sufficiently small

$$x_0 := \max \left\{ \hat{x} \in [0,1] \mid b(\lambda, u_n(\hat{x}), \hat{x}) = 0 \text{ and } b(\lambda, u_n(x), x)_{|(\hat{x}-\delta,\hat{x})} > 0 \right\}. \quad (3.58)$$

Note, that by our assumptions x_0 is independent from $n \in \mathbb{N}$ for fixed λ and b changes sign at $x = x_0$ by (3.58).
Furthermore there exists possibly a unique point $x_s = x_s(n)$ defined by

$$x_s(n) := \left\{ \tilde{x} \in [0,1] \mid \int_{\tilde{x}}^1 b(\lambda, u_n(\tau), \tau)d\tau = 0 \text{ and} \right.$$
$$\left. \int_x^1 b(\lambda, u_n(\tau), \tau)d\tau > 0 \text{ for every } x < \tilde{x} \right\}. \quad (3.59)$$

Note, that existence of $x_s = x_s(n)$ is by no means guaranteed. But if x_s exists, then $x_s < x_0$ holds. Furthermore we define

$$\tilde{x}_s(n) := \begin{cases} x_s(n) & \text{, if } x_s \text{ exists} \\ 0 & \text{, elsewhere.} \end{cases}$$

Due to the obvious fact that $\tilde{x}_s(n) < x_0$ there exists $\varrho(n) > 0$ with $\tilde{x}_s(n) + \varrho(n) = x_0$. We define

$$\varrho_0 := \inf_{n \in \mathbb{N}} \{\varrho_n\}$$

and it is easy to see that $\varrho_0 > 0$ is valid.
With these definitions at hand we can prove the following:

Lemma 44. Let $z_\varepsilon \in \mathcal{C}_\varepsilon^+$. We claim: For every $\varrho \in (0, \varrho_0)$ exists $\varepsilon_0 = \varepsilon_0(\varrho)$ such that for all $\varepsilon < \varepsilon_0$ we obtain

$$z'_{\varepsilon|(0,x_0-\varrho)} < 0.$$

Proof. We argue by contradiction and thus we assume the existence of $\varrho_1 \in (0, \varrho_0)$ such that for every $\varepsilon_0 > 0$ there exist $\varepsilon < \varepsilon_0$ and $x_\varepsilon \in (0, x_0 - \varrho_1)$ such that $z'_\varepsilon(x_\varepsilon) \geq 0$. By virtue of Theorem 35 and Lemma 43 we obtain a sequence $(z_n)_{n \in \mathbb{N}} \in \mathcal{C}^+_{\varepsilon_n}$ with the following property: The function z_n changes sign exactly once at $x_1 = x_1(n)$ and moreover we have

$$z'_{n|(0,x_1(n))} < 0 \quad \text{and} \quad z'_{n|(x_1(n),1)} > 0.$$

Due to our assumption we deduce $z'_{n|[x_0-\varrho_1,1]} > 0$ for every $n \in \mathbb{N}$. Helly's theorem yields $z_{n|[x_0-\varrho_1,1]} \rightarrow z_{|[x_0-\varrho_1,1]}$ pointwise and in particular

$$z_{|[x_0-\varrho_1,1]} \text{ is monotonically increasing.} \tag{3.60}$$

With respect to the definition of x_s and ϱ_0 we obtain for every $x \in [x_0 - \varrho_1, x_0]$

$$\sigma(z(x)) = \int_x^1 b(\lambda, u(\tau), \tau) d\tau \leq 0, \tag{3.61}$$

where $u(x) = \int_0^x z(s) ds$, and hence

$$z_{|[x_0-\varrho_1,x_0]} \leq 0. \tag{3.62}$$

Furthermore we know that the map $x \mapsto \int_x^1 b(\lambda, u(\tau), \tau) d\tau$ is monotonically decreasing for $x \in (x_0 - \varrho_1, x_0)$. Thus (3.60) and (3.61) yields one of the two possibilities:

(i) Either $z_{|[x_0-\varrho_1,x_0]} \in (\nu_1, \nu_2) \subseteq \mathbb{R}^+$

(ii) or $z_{|[x_0-\varrho_1,x_0]} \equiv \text{const.}$.

Part (i) contradicts (3.62) and part (ii) implies

$$\int_x^1 b(\lambda, u(\tau), \tau) d\tau = \sigma(z(x)) = \text{const.}$$

for every $x \in [x_0 - \varrho_1, x_0]$ contradicting (3.58). $\qquad\square$

Let $(u_n)_{n \in \mathbb{N}} \in \mathcal{C}^+_{\varepsilon_n}$ be a sequence for fixed $\lambda > 0$ with $\varepsilon_n \searrow 0$ and by virtue of Theorem 41 we obtain functions u, z such that $\lim_{n \to \infty} u_n =: u$ and $\lim_{n \to \infty} z_n =: z = u'$. We define

$$A := \left\{ x \in [0,1] \mid \int_x^1 b(\lambda, u(\tau), \tau) d\tau \in \mathbb{R}\backslash[\sigma(\nu_a), \sigma(\nu_b)] \right\}$$

and due to the identity

$$\sigma(z(x)) = \int_x^1 b(\lambda, u(\tau), \tau) d\tau \tag{3.63}$$

we have $A = \{ x \in [0,1] \mid z(x) \in \mathbb{R}\backslash[\nu_a, \nu_b] \}$. (Recall for a definition of ν_a, ν_b (3.8).) Also by (3.63) we deduce $z_{|A}$ is continuously differentiable. Furthermore we define

$$B := [0,1]\backslash A$$

and we claim $B \subseteq [0, x_0 - \varrho_0]$: Let $x \in B$ and by definition this implies

$$\int_x^1 b(\lambda, u(\tau), \tau) d\tau \in (\sigma(\nu_a), \sigma(\nu_b)) \subseteq \mathbb{R}^+.$$

Therefore we have

$$\int_x^1 b(\lambda, u_n(\tau), \tau) d\tau \geq 0$$

for $n > n_0$ and hence by (3.59) we obtain

$$x < x_s(n) = x_0 - \varrho(n) < x_0 - \varrho_0.$$

For the sequel of the chapter we assume the existence of $\lambda \in \mathbb{R}^+$ and u obtained by the usual limit process such that

$$\int_{x_2}^1 b(\lambda, u(\tau), \tau) d\tau \in B \quad \text{for a certain } x_2 \in [0, x_0].$$

In particular we have $\int_0^1 b(\lambda, u(\tau), \tau) d\tau > 0$. Thus $x_s(n) \in (0, 1)$ defined in (3.59) exists without loss of generality for every $n \in \mathbb{N}$ and by virtue of Lemma 44 we have

$$z'_{n|(0, x_0 - \frac{\varrho_0}{2})} < 0$$

for all $n \in \mathbb{N}$. By Helly's theorem we obtain

$$z_{|(0, x_0 - \frac{\varrho_0}{2})} \text{ is monotonically decreasing,}$$

where z denotes the pointwise limit of $(z_n)_{n \in \mathbb{N}}$.

Because of $B \subseteq [0, x_0 - \frac{\varrho_0}{2})$ we immediately deduce, that $z_{|B}$ is monotonically decreasing and in particular the function has at most finitely many jumps in B. Moreover we have

$$\int_x^1 b(\lambda, u(\tau), \tau) d\tau_{|B} = \sigma(z)_{|B} \text{ is monotonically decreasing}$$

and due to the shape of σ we obtain the following (recall that the Maxwell points of W are denoted by ν_m, ν_M):

Theorem 45. *Let $(z_n)_{n \in \mathbb{N}}$ be a sequence with $z_n \in \mathcal{C}_{\varepsilon_n}^+$ for every $n \in \mathbb{N}$ and for fixed $\lambda > 0$ and let z be the pointwise limit of the sequence. Furthermore let $u(x) := \int_0^x z(s) ds$. Then z has at most one jump in $(0, 1)$ in accordance with the first and the second Weierstrass–Erdmann corner condition.*
Furthermore we have the following: If $\int_0^1 b(\lambda, u(\tau), \tau) d\tau < \sigma(\nu_m)$, then z has no jump and if $\int_0^1 b(\lambda, u(\tau), \tau) d\tau > \sigma(\nu_1)$, then z suffers exactly one jump discontinuity at a not necessarily unique point x_J satisfying

$$\int_{x_J}^1 b(\lambda, u(\tau), \tau) d\tau = \sigma(\nu_m). \tag{3.64}$$

Moreover we obtain $z(x_J - 0) = \nu_M$ and $z(x_J + 0) = \nu_m$.
If $\int_0^1 b(\lambda, u(\tau), \tau) d\tau \in [\sigma(\nu_m), \sigma(\nu_1)]$, then z has either no jump or exactly one at $x = x_J$ such that again (3.64) holds.

Proof. The proof is obvious due to the fact that z satisfies the first and the second Weierstrass–Erdmann corner condition and moreover z is monotonically decreasing in B. Note, that z cannot jump in the region where z is monotonically increasing, because z satisfies the corner conditions and

$$\sigma(z(x)) = \int_x^1 b(\lambda, u(\tau), \tau) d\tau \le 0 < \sigma(\nu_m)$$

for every $x \in [x_0 - \varrho_0, 1]$. $\qquad\square$

By Theorem 45 we obtain the following shape for the deformation of the bar dependening on the body force:

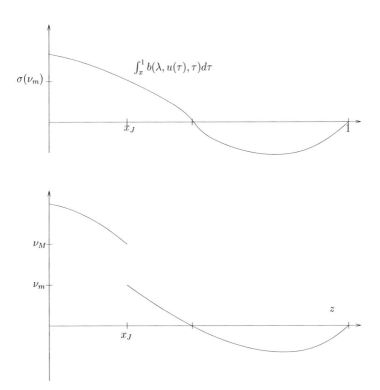

For the sequel of the section we consider body forces independent of u (such a force is commonly called dead load) and with certain symmetries to prove an analogues statement as in Theorem 45. In particular, we consider a body force with

$$b\left(\lambda, \frac{1}{2} - x\right) = b\left(\lambda, \frac{1}{2} + x\right)$$

for every $x \in [0, \frac{1}{2}]$ and furthermore we assume the existence of $x_0 \in (0, \frac{1}{2})$ such that

$$b_{|(0,x_0)} \geq 0 \quad \text{and} \quad b_{|(x_0,\frac{1}{2})} \leq 0.$$

Moreover we assume existence of open subsets $\Omega_1 \subseteq [0, x_0)$ and $\Omega_2 \subseteq (x_0, \frac{1}{2}]$ with $|\Omega_j| > 0$ for $j = 1, 2$ such that $b_{|\Omega_1} > 0$ and $b_{|\Omega_2} < 0$. We assume also existence of $\vartheta > 0$ with $b(\lambda, u, \cdot)_{|(0,\vartheta)} > 0$ for every $\lambda \in \mathbb{R}^+$ and $u \in \mathbb{R}$.

By a similar procedure like above applied to the interval $[0, \frac{1}{2}]$ respectively $[\frac{1}{2}, 1]$ one can deduce by reflection of $z_{|[0,\frac{1}{2}]}$ to the unit interval the following:

Theorem 46. *There exists an unbounded branch $\mathcal{C}_\varepsilon^+$ of solutions of (3.5) having projection $[0, \infty)$ on the λ–axis. Let $z \in \mathcal{C}_\varepsilon^+$, then we have:*

(i) z satisfies $z\left(\frac{1}{2} - x\right) = z\left(x + \frac{1}{2}\right)$ for every $x \in [0, \frac{1}{2}]$.

(ii) z' has at most 3 zeros in $(0, 1)$ and in particular we have $z'\left(\frac{1}{2}\right) = 0$.

Again by the same techniques we prove:

Theorem 47. *Let $z_n \in \mathcal{C}_{\varepsilon_n}^+$ be a sequence for fixed $\lambda > 0$ and for $\varepsilon_n \searrow 0$. Then we obtain*

(i)

$$z_n \to z \quad \text{pointwise in } [0, 1]. \tag{3.65}$$

(ii) The limit z solves the equation

$$\sigma(z(x)) = \int_x^1 b(\lambda, \tau) d\tau \tag{3.66}$$

for every $x \in [0, 1]$

(iii) and z satisfies the first and the second Weierstrass–Erdmann corner conditions.

By virtue of Theorem 47 we get:

Theorem 48. *Let z be a function obtained by the limiting process (3.65). If*

$$\int_{\frac{1}{2}}^1 b(\lambda, \tau) d\tau < \sigma(\nu_m),$$

then z has no jump and if

$$\int_{\frac{1}{2}}^1 b(\lambda, \tau) d\tau > \sigma(\nu_1),$$

then z suffers exactly two jump discontinuities at not necessarily unique points x_{J_1} and x_{J_2} satisfying

$$\int_{x_{J_j}}^1 b(\lambda, \tau) d\tau = \sigma(\nu_m) \tag{3.67}$$

for $j = 1, 2$. Moreover we obtain $z(x_{J_1} - 0) = \nu_m$, $z(x_{J_1} + 0) = \nu_M$ and $z(x_{J_2} - 0) = \nu_M$, $z(x_{J_2} + 0) = \nu_m$.

If $\int_{\frac{1}{2}}^1 b(\lambda, \tau) d\tau \in [\sigma(\nu_m), \sigma(\nu_1)]$, then we have either no jump or exactly two jumps at x_{J_1} and x_{J_2} such that (3.67) holds.

3.7   The Limiting set is a continuum

Let $(z_n)_{n \in \mathbb{N}}$ be a sequence with $z_n \in \mathcal{C}_{\varepsilon_n}^+$ for every $n \in \mathbb{N}$ and for fixed $\lambda \geq 0$. By virtue of Lemma 22 we obtain a constant $K > 0$ such that

$$||z_n||_\infty \leq K \quad \text{and} \quad ||u_n||_{C^1} \leq 2K. \tag{3.68}$$

We define for some $q < \infty$ the set

$$\Sigma_0^+ = \Big\{ (\lambda, u) \in \mathbb{R} \times W^{1,q}(0,1) \mid (\lambda_n, u_n, v_n, \mu_n) \in \bar{\mathcal{C}}_{\varepsilon_n}^+, \lambda_n \to \lambda, u_n \to u \tag{3.69}$$
$$\text{in } W^{1,q}(0,1) \text{ as } \varepsilon_n \searrow 0 \Big\}.$$

Furthermore we introduce the solution set

$$\Sigma_\varepsilon^+ = \big\{ (\lambda, u) \mid (\lambda, u, v, \mu) \in \bar{\mathcal{C}}_\varepsilon^+ \big\}.$$

Because $\mathcal{C}_\varepsilon^+$ forms a continuum the set Σ_ε^+ is connected and compact.
Let $B_r \subseteq W^{1,q}(0,1)$ denote the closed ball of radius r, centered at the origin. For any number $\lambda_0 > 0$ we consider the sets

$$A = \{0\} \times B_{K_1} \quad \text{and} \quad B = \{\lambda_0\} \times B_{K_2}$$

with constants K_1 and K_2 independent of ε such that $(3.68)_2$ holds. Moreover we introduce the subset
$$\Upsilon_{\varepsilon,\lambda_0}^+ := \Sigma_\varepsilon^+ \cap \big([0,\lambda_0] \times W^{1,q}(0,1) \big)$$

and we have

$$A \cap \Upsilon_{\varepsilon_n,\lambda_0}^+ \neq \emptyset \text{ and } B \cap \Upsilon_{\varepsilon_n,\lambda_0}^+ \neq \emptyset$$
$$\text{are not separated in } \Upsilon_{\varepsilon_n,\lambda_0}^+ \text{ for any } \varepsilon_n > 0.$$

For a definition of separation of two sets see [1], section 1. To apply the main theorem of [1], section 3, we need the following:

Lemma 49. *For each neighborhood N of $\Sigma_0^+ \cap ([0,\lambda_0] \times W^{1,q}(0,1))$, there exists $n \in \mathbb{N}$ such that $\Upsilon_{\varepsilon_n,\lambda_0}^+ \subseteq N$.*

Proof. We argue by contradiction. Thus we assume existence of a neighborhood N_0 of $\Sigma_0^+ \cap ([0,\lambda_0] \times W^{1,q}(0,1))$ such that $(\lambda_n, u_n) \in \Upsilon_{\varepsilon_n,\lambda_0}^+$ and $(\lambda_n, u_n) \notin N_0$ for every $n \in \mathbb{N}$. For some subsequence we obtain $\lambda_n \to \lambda \in [0,\lambda_0]$.
Furthermore we claim $u_n \to u$ in $W^{1,q}(0,1)$: By our growth conditions on b and σ and the bound on $(\lambda_n)_{n \in \mathbb{N}}$ we obtain as in the proof of Theorem 19 the bound $||z_n||_{L^{p+1}} \leq C$ with C independent of $n \in \mathbb{N}$. In the same way as in Lemma 22 one can prove

$$||z_n||_{C^0} \leq C$$

and Theorem 35 yields by Helly's theorem and the well known theorem of Radon–Riesz $u_n \to u$ in $W^{1,q}(0,1)$. In particular, $(\lambda, u) \in \Sigma_0^+ \cap ([0,\lambda_0] \times W^{1,q}(0,1))$, i.e. $(\lambda_n, u_n) \in N_0$ for n sufficiently large, a contradiction. $\qquad\square$

In view of [1] we deduce

$$A \cap \left(\Sigma_0^+ \cap \left([0, \lambda_0] \times W^{1,q}(0,1) \right) \right) \text{ and } B \cap \left(\Sigma_0^+ \cap \left([0, \lambda_0] \times W^{1,q}(0,1) \right) \right)$$

are not separated in $\Sigma_0^+ \cap ([0, \lambda_0] \times W^{1,q}(0,1))$. By Corollary 4 in [1] we may conclude since λ_0 was chosen arbitrarily:

Theorem 50. *The set* $\Sigma_0^+ \subseteq \mathbb{R} \times W^{1,q}(0,1)$ *is a continuum having projection* $[0, \infty)$ *on the parameter axis.*

The limiting continuum especially for body forces like $b(\lambda, u, x) = \lambda g(x)$ with $\int_0^1 g(\tau) d\tau > 0$ and where g satisfies (H-1)– (H-4) can be characterized in the following way:

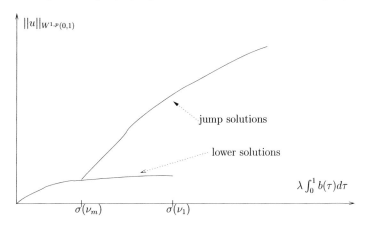

Note, that this picture is an easy consequence of Theorem 45. By "lower solutions" we mean those solutions of (3.63) where no jump occurs.

3.8 Minimizing properties of solutions for dead loading

We consider a body force $b = b(\lambda, x)$ independent of the placement u and let z_* be the jump solution obtained by the singular limit process, if $\int_0^1 b(\lambda, \tau) d\tau > \sigma(\nu_m)$ and a lower solution otherwise. Then we deduce the following:

Theorem 51. *Let* (λ, z_*) *be as described above. Then* $J(\lambda, u)$ *defined by* (3.1) *attains its minimum at* $u_* := \int_0^x z_*(s) ds$, *i.e.*

$$\min_u \left\{ J(\lambda, u) \mid u \in W^{1,p+1}(0,1),\ u(0) = 0 \right\} = J(\lambda, u_*).$$

We omit the proof because it is completely analogues to the proof of Theorem 6.4 given in [18].

Chapter 4

Stability analysis

4.1 Introduction

In this chapter we consider a homogenous elastic bar placed in a soft loading device with body–force potential bu, which determines a certain number of different phases in the bar. The total energy of the bar is given by

$$J(u) := \int_0^1 W(u') - bu\, dx - u(1)t_r,$$
$$u(0) = 0. \tag{4.1}$$

The boundary condition is due to the fact that the bar is fixed at its left end–point. Furthermore the function $x \mapsto u(x)$ is the placement of the bar, $x \mapsto u'(x)$ is the deformation, we call $x \mapsto u'(x) - 1$ the strain and $x \mapsto W'(u'(x))$ the stress of the bar. Moreover the soft loading device produces a load at $x = 1$ given by a certain constant t_r. Our goal is to obtain properties of stable solutions of the Euler–Lagrange equation of the corresponding singular perturbed variational problem to (4.1).

A variational problem like (4.1) was considered by a lot of authors and the list of them we give is by no means complete. First we mention [15], who solves the problem with body force 0, and [23], who considers a physically more reasonable inhomogenous bar in a soft respectively hard loading device and gives necessary and sufficient conditions for the existence of a weak relative minimum of the related variational problem.

However, our approach is totally different, because we propose certain body forces, such that the existence of a global minimizer of (4.1) is trivial by the direct methods of the calculus of variations and moreover we determine the shape of the minimizer. Note, that existence of a global minimizer of the variational problem is well known by [10], because (4.1) is equivalent to

$$\int_0^1 (W(u') - t_r u' - bu)\, dx := \int_0^1 \hat{W}(u') - bu\, dx$$

and the last term is obviously concave in u.

In fact, problems like (4.1) are not very realistic even in cases where microstructures are observed, because energy minimization determines usually infinitesimal fine mixture of phases, whereas microstructure in nature is of finite size. To determine the length scale it is important to introduce an additional term in the energy, a so called interfacial energy term controlled by a parameter $\varepsilon > 0$ and ε is assumed to be small. The singular perturbed functional reads as follows:

$$J_\varepsilon(u) := \int_0^1 \left(\frac{\varepsilon}{2} u''^2 + W(u') - b(x)u \right) dx - u(1)t_r, \tag{4.2}$$
$$u(0) = 0.$$

The interfacial energy causes the deformation of the bar to be continuous, but it allows the deformation to suffer rapid changes over small intervals. This approach was also fruitful in detecting some length scales in microstructures especially in the one–dimensional problem

$$J_\varepsilon(u) := \int_0^1 \varepsilon^2 u''^2 + (u'^2 - 1)^2 + u^2 \, dx \tag{4.3}$$

subject to periodic boundary conditions. The main result in [32] is, that every minimizer of (4.3) is periodic with minimal period of order $\varepsilon^{\frac{1}{3}}$. But also in higher dimensional problems an interfacial energy was used to obtain certain length scales depending on ε, see for example [29].

Our aim is to analyze the stable configurations of the bar and moreover we want to discover special features of stable solutions of the corresponding Euler–Lagrange equation to (4.2) such as the width of an interval in which the deformation changes from one phase to another. To do this we propose a similar approach like in [2], [3] and [4]. They also obtain certain structures of solutions of the Euler–Lagrange equation by virtue of positive definiteness of the second variation. However, in [3] and [4] neither body–force nor an additional traction is included in their model. In [2] they analyze the phase field model and the corresponding total energy can be written in the following way, where F denotes a double–well potential:

$$\int_\Omega \left(\frac{\varepsilon^2}{2} |\nabla u|^2 + F(u) - \varepsilon T(x)u \right) dx.$$

This model has the nice property, that as ε tends to zero the last term becomes more or less absent and therefore it causes just the number of phase transitions but not the placement of the bar within one phase. Therefore it seems to be surprising to us, that our model yields similar results as in the phase–field model.

The plan of this chapter is as follows:
In the first section we discuss briefly the physical meaning of the proposed variational problem. To this purpose we follow the lines of [23]. Due to Helly's theorem applied to functions with bounded variation we can prove in the second section the existence of a limit in a certain function–space as the perturbation parameter ε tends to zero. Moreover we can show in which way the number of phase transitions in the perturbed variational problem depends on the body force and the traction. In the third section we will determine the shape of the minimizer or, more generally, of a stable critical point of the perturbed problem. To prove these results we proceed in a similar way as in [4] and especially in [2]. Nevertheless it seems to be very interesting to us that we can actually

prove uniform convergence of $(u'_\varepsilon)_\varepsilon$ to the global minimizer of the unperturbed problem in certain phase regions for every sequence of stable critical points of the perturbed problem under some conditions on the body–force b and the traction. This result seems to be new in the context of singular perturbations of nonconvex variational problems, because in [2] a similar statement was only given for the sequence of global minimizers of the perturbed variational problem. In particular, no argument is used in our proof which depends on the energy. Furthermore a convergence–rate of $(u'_\varepsilon)_\varepsilon$ with respect to ε is given in a similar way as in [2]. In the last section we obtain a necessary condition for uniqueness of the global minimizer of (4.2) for ε sufficiently small . For this we assume essentially a gravitationally body force, i.e. $b(x) = \varrho(x)g$. Thereby g denotes a non–negative constant and ϱ is a function bounded away from zero. To prove this we borrow heavily on a procedure proposed in [6]. In their work they were able to show, that between an upper and a lower solution of the Euler–Lagrange equation there is exactly one solution by rewriting the static Euler–Lagrange equation as a dynamical system and by proving that every equilibrium lying between the upper and the lower solution is asymptotically stable. The same ideas were also used in [4]. Moreover we mention that a uniqueness result for the global minimizer of the relaxed variational problem was obtained in [9] without body force and traction, but with the constraint, that the bar has a certain length in the deformed configuration. However, our result seems to be new if body–force and traction is present.

4.1.1 The model

We consider a one dimensional homogenous elastic bar and without loss of generality we assume the length of the undeformed bar to be one. Let $[0,1]$ be the bar interval. The function $x \mapsto u(x)$ is the placement of the bar, $x \mapsto u'(x)$ is the deformation (provided the derivative exists in a weak sense), we call $x \mapsto u'(x) - 1$ the strain. Let $W \in C^2(\mathbb{R})$ be the stored–energy density of the bar and $x \mapsto W'(u'(x)) := \sigma(u'(x))$ is called the stress. Since the bar is assumed to be homogenous, W is independent of the spatial variable and furthermore it is a typical double–well potential, that means there exists $\nu_1 < \nu_2$ such that

$$W'_{|(-\infty,\nu_1)} \quad \text{is strictly increasing,}$$
$$W'_{|(\nu_1,\nu_2)} \quad \text{is strictly decreasing,}$$
$$W'_{|(\nu_2,\infty)} \quad \text{is strictly increasing.}$$

In additional we assume polynomial growth for W, namely:

$$c_1 + c_2 |F|^p \leq W(F) \leq c_3 + c_4 |F|^p,$$

$c_1, \ldots, c_4$ are some constants with $c_j > 0$ for $j = 2,3,4$ and $p \geq 2$. In particular we assume

$$||\sigma(\nu)\nu|| \leq c_5 + c_6 ||\nu||^p \tag{4.4}$$

with constants $c_5, c_6 > 0$ and

$$\lim_{\nu \to \pm\infty} \frac{\sigma(\nu)\nu}{||\nu||^p} \geq K > 0. \tag{4.5}$$

Furthermore let ν_m and ν_M be defined by the Maxwell conditions.
In our model a body force potential B linear in the placement is present, that means

$$B(u, x) = -b(x)u,$$

and the body force b is some real–valued continuous differentiable function. Further assumptions on the body force will be made in the forthcoming sections.
The left end of the bar should always be held fixed and due to that we get

$$u(0) = 0.$$

The right hand side is assumed to be placed in a soft loading device which produces an additional traction t_r. The total energy of the bar is defined by

$$J(u) := \int_0^1 \left(W(u') - b(x)u \right) dx - t_r u(1),$$

$$u(0) = 0. \tag{4.6}$$

Usually global minimizers (existence provided) or stable critical points of this variational problem have jumps in the deformation caused by the loss of convexity of W. We will say that the bar changes its phase at a point where u' is discontinuous. In our model the change is produced by deformation of the bar alone.
Our approach is to plug in an additional energy term in the variational problem, the interfacial energy between phases. This procedure assigns the problem (P_ε) which will be discussed in the forthcoming sections:

$$J_\varepsilon(u) := \int_0^1 \left(\frac{\varepsilon}{2} u''^2 + W(u') - b(x)u \right) dx - t_r u(1),$$

$$u(0) = 0,$$

where $\varepsilon > 0$ denotes a small parameter. Considering an interfacial energy is a standard tool to ensure the existence of a global minimizer by the direct methods of the calculus of variations and in order to make the corresponding Euler–Lagrange equation elliptic, which causes much higher regularity of some critical point. In particular, we will have no jump in the deformation but rather rapid changes within a small interval indicating a change of phase for $\varepsilon = 0$.

4.2 Existence of a minimizer for the unrelaxed problem

As denoted in the introduction we consider the following unperturbed variational problem

$$J(u) := \int_0^1 W(u'(x)) - b(x)u \, dx - t_r u(1),$$

$$u(0) = 0. \tag{P}$$

Let $\sigma = W'$ the stress and ν_m, ν_M the Maxwell points of W such that $\nu_m < \nu_M$. In particular we know by the Maxwell condition the relation $\sigma(\nu_m) = \sigma(\nu_M)$. Let

$\nu_1 < \nu_2$ these points where W looses convexity and furthermore there exists ν_a, ν_b, both uniquely defined by

$$\begin{aligned} \nu_a &:= \min \left\{ x \mid x = \sigma^{-1}(\nu_2) \right\}, \\ \nu_b &:= \max \left\{ x \mid x = \sigma^{-1}(\nu_1) \right\}, \end{aligned} \tag{4.7}$$

and it is obvious from the definition that one obtains $\nu_a < \nu_b$. The typical shape of σ described here can be seen in the following picture:

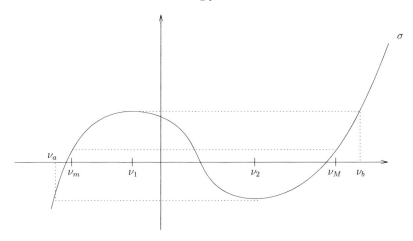

Furthermore we impose the following condition on b:

$$\begin{aligned} &\text{The function } b \text{ possesses exactly } s \text{ exclusively} \\ &\text{non--degenerate zeros for a certain } s \in \mathbb{N}. \end{aligned} \tag{4.8}$$

Definition 52. We define the set

$$X := \left\{ x \in (0,1) \mid x \mapsto \int_x^1 b(\tau)d\tau + t_r - \sigma(\nu_m) \text{ changes sign} \right\} \tag{4.9}$$

and because of (4.8) there exists discrete points $x_1, \ldots, x_r$ for a certain $r \in \mathbb{N}$, $r \le s$, with $X = \{x_1, \ldots, x_r\}$.
Furthermore we define for $j = 1, \ldots, r$ the set

$$A_j := \left\{ x \in [0,1] \mid x \mapsto \int_x^1 b(\tau)d\tau + t_r - \sigma(\nu_m) \quad \text{is monotonic and } x_j \in A_j \right\}. \tag{4.10}$$

Note, that by (4.9) we have $x_j \in \text{int}(A_j)$.

With respect to this definition each A_j is a closed interval in $[0,1]$ and we will denote it by $[x_1^j, x_2^j]$ for $j = 1, \ldots, r$.
Moreover we assign another hypothesis on b:

(H–1) Let

$$Y := \left\{ x \in [0,1] \mid \int_x^1 b(\tau)d\tau + t_r - \sigma(\nu_m) = 0 \right\} \qquad (4.11)$$

and we assume

$$Y = X,$$

where X is defined by (4.9). In particular, let $y \in Y$, then we have $y = x_j \in$ int(A_j) for some $j = 1, \dots, r$.

First we address the problem of existence of a global minimizer of (P) in a certain function space. Therefore we consider the relaxed variational problem

$$J^c(u) := \int_0^1 W^c(u') - b(x)u \, dx - t_r u(1), \qquad \text{(RP)}$$
$$u(0) = 0$$

and our goal is to solve

$$\min \left\{ J^c(u) \mid u \in W^{1,p}(0,1) \text{ with } u(0) = 0 \right\}. \qquad (4.12)$$

Existence of a minimizer of Problem (4.12) is easily proved by the direct methods of the calculus of variations (see for example [11]). This minimizer satisfies

$$\frac{d}{dx}\sigma^c(u') = -b$$

pointwise almost everywhere and the boundary conditions

$$u(0) = 0, \ \sigma^c(u'(1)) = t_r.$$

Here, $\sigma^c := (W^c)'$. Integration of this equation yields

$$\sigma^c(u'(x)) = \int_x^1 b(\tau)d\tau + t_r, \qquad (4.13)$$
$$u(0) = 0.$$

In the region where σ^c is invertible, we have the following:

$$z(x) := u'(x) = (\sigma^c)^{-1}\left(\int_x^1 b(\tau)d\tau + t_r \right). \qquad (4.14)$$

Note, that equation (4.14) determines the minimizer for every $x \in [0,1] \setminus X$ by (H–1). For $x = x_1, \dots, x_r$ the function z has to jump from ν_m to ν_M or vice versa. Hence, z given by (4.14) is obviously the only solution of (4.13) and we obtain $u(x) := \int_0^x z(s)ds$ is the unique minimizer of (RP) with

$$\text{supp}(z) \subseteq \{x \in \mathbb{R} \mid W(x) = W^c(x)\}.$$

Moreover, we assume b and σ as smooth as necessary such that $z \in C^2((0,1)\setminus X)$. Furthermore the direct methods of the calculus of variations yield $\inf(P) = \min(RP)$ and this implies the following:

Lemma 53. *There exists a unique global minimizer $u \in C^3((0,1)\backslash X)$ of (P). Furthermore formula (4.14) holds for u' and for every $x \in X$ the deformation u' has to jump from ν_m to ν_M or vice versa.*

Proof. It remains to show uniqueness: We assume for a moment there exists at least two global minimizers u_1 and u_2 of (P). Because of $\inf(P) = \min(RP)$ we know, that u_1 and u_2 are also global minimizers of (RP), a contradiction to the above mentioned fact that the solution of (4.13) is unique. □

In the sequel of the chapter we analyze the variational problem depending on the capillarity coefficient $\varepsilon > 0$, namely

$$J_\varepsilon(u) := \int_0^1 \frac{\varepsilon}{2}u''^2 + W(u') - b(x)u\, dx - t_r u(1), \tag{P_ε}$$
$$u(0) = 0,$$

defined over the function space $\mathcal{A} := H^2(0,1) \cap \{v \mid v(0) = 0\}$. Critical points of (P_ε) satisfy

$$-\varepsilon u^{(4)} + \frac{d}{dx}[W'(u')] + b = 0, \ 0 < x < 1, \tag{4.15}$$

with natural boundary conditions

$$u''(0) = u''(1) = 0, \qquad \varepsilon u'''(1) = W'(u'(1)) - t_r. \tag{4.16}$$

Integration of (4.15) yields the system

$$u' = z,$$
$$-\varepsilon z'' + \sigma(z) = \int_x^1 b(\tau)d\tau + t_r, \tag{4.17}$$
$$u(0) = z'(0) = z'(1) = 0.$$

Note, that every solution of (4.17) satisfies $z \in C^3(0,1)$ by standard regularity theory. Our approach is based on studying stable solutions of (4.17).

Definition 54. A solution $z_\varepsilon = u'_\varepsilon$ of (4.17) is called stable, if the second variation of $J_\varepsilon(u_\varepsilon)$ defines a positive quadratic form over the function space $\mathcal{A} \times \mathcal{A}$.

In the sequel we consider a sequence $(z_n)_{n \in \mathbb{N}}$ of stable critical points of (P_{ε_n}) with the property $\varepsilon_n \searrow 0$ for $n \to \infty$. We address the following two basic problems:

1. Does the sequence $(z_n)_{n \in \mathbb{N}}$ converge to a function z in a certain function space?

2. Which properties has the limit z and moreover what are the conditions to ensure z to be the global minimizer of (P)?

4.3   Compactness of a sequence of stable critical points

In this section we are interested in the topic whether an arbitrary sequence $(u_n)_{n \in \mathbb{N}}$ of stable critical points of (P_{ε_n}) converges in any sense to a function u. Another goal is to determine u as the minimizer of (P) provided some properties of $(u_n)_{n \in \mathbb{N}}$ respectively b are fulfilled. To obtain a result we follow the lines of [2] and we start with the key observation which is essentially step 1 of Lemma 2.1 in the above mentioned paper. Nevertheless we give the proof for the readers convenience and because of the fact it is slightly different to the one stated in [2].

Lemma 55. *Let z be a stable solution of* (4.17) *for some $\varepsilon > 0$ in the sense described in Definition 54. Then there exists no nodal domain Ω of z' such that* $\text{sign}(z'b)_{|\Omega} > 0$.

Proof. We prove by contradiction and thus we assume the existence of a nodal domain $[c,d]$ of z' with the property $z'(x)b(x) > 0$ for every $x \in [c,d]$, a situation like in the following picture:

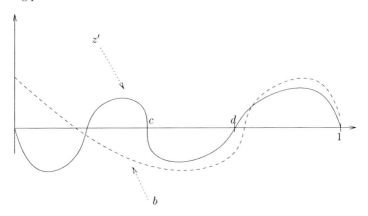

Differentiating (4.17) and multiplying the equation by z' yields

$$\varepsilon z''' z' - \sigma'(z) z'^2 = b z',$$
$$z'(c) = z'(d) = 0. \tag{4.18}$$

Furthermore we define

$$v'(x) := \begin{cases} z'(x), & \text{if } x \in [c,d] \\ 0, & \text{elsewhere.} \end{cases}$$

Thus v' is continuous on $(0,1)$ and moreover we define $v(x) := \int_0^x v'(s)ds$ and it is easy to see that $v \in \mathcal{A}$ and in particular v is a suitable test function of (P_ε). Due to (4.18) and integration by parts we get

$$\int_c^d \varepsilon z''' z' - \sigma'(z) z'^2 dx = \int_c^d -\varepsilon z''^2 - \sigma'(z) z'^2 dx = \int_c^d z' b \, dx > 0. \tag{4.19}$$

On the other hand the inequality in (4.19) yields

$$D^2 J_\varepsilon(u)(v,v) = \int_0^1 \varepsilon v''^2 + \sigma'(z)v'^2 dx = \int_c^d \varepsilon z''^2 + \sigma'(z)z'^2 dx < 0 \qquad (4.20)$$

contradicting the assumed stability of z. □

Remark 56. Note, that we cannot prove Lemma 55 in the same way as above, if the bar is placed in a hard loading device (that means, that the bar is fixed on the left and additionally on the right hand side). The problem depends on the fact that the above constructed v is not an allowed test function in the hard loading device.

An immediate consequence of Lemma 55 is the following:

Corollary 57. *We assume that b changes sign in $(0,1)$ exactly s times and let u be a stable critical point of (P_ε). Then $z' = u''$ has at most s changes of sign in $(0,1)$.*

For the proof we refer the reader to [2], Step $2 - 4$ of Lemma 2.1.
By the above Corollary we can prove the following convergence result:

Lemma 58. *Let z_ε be a stable solution of (4.17) and let $u_\varepsilon(x) := \int_0^x z_\varepsilon(s)ds$. Then for any sequence $(\varepsilon_n)_{n\in\mathbb{N}}$ approaching zero there exists a subsequence of corresponding u_ε with the following properties:*

(i) We have $u_\varepsilon \to u$ uniformly

(ii) and $z_\varepsilon \to z$ pointwise and $z \in L^\infty(0,1) \cap BV(0,1)$.

(iii) The limit z satisfies the equation

$$\sigma(z(x)) = \int_x^1 b(\tau)d\tau + t_r \qquad (4.21)$$

for every $x \in [0,1]$

(iv) and z fulfills the first and the second Weierstrass–Erdmann corner conditions.

Proof. Completely identical to the proofs of Chapter 3 one can show

$$||z_n||_\infty \leq K \qquad (4.22)$$

with a constant K independent of $n \in \mathbb{N}$. This yields (i) by compact embedding $C^1([0,1]) \hookrightarrow C^0([0,1])$. By the fundamental theorem in calculus and Corollary 57 we obtain, that the total variation of $(z_n)_{n\in\mathbb{N}}$ is also uniformly bounded and hence Helly's theorem (see for example [34]) implies (ii). The last two items are proved like the corresponding results in the previous chapter. □

We continue with the discussion of some properties of the above mentioned limit.

Lemma 59. *Let $(z_n)_{n\in\mathbb{N}}$ be a sequence of stable critical points of (4.17) such that $z_n \to z$ pointwise for $\varepsilon_n \searrow 0$. Then the following statements are valid:*

(i) For every $x \in [0,1]$ we have $z(x) \in (-\infty, \nu_1] \cup [\nu_2, \infty)$.

(ii) Let $(z_n)_{n\in\mathbb{N}}$ be a sequence of global minimizers of (P_{ε_n}), then z is the (unique) global minimizer of (P).

In any case, the limit is stable because it attains no values in the spinodal region.

Proof. ad(i): Let z be the limit of stable solutions of (4.17) in the way described above and suppose the given statement is wrong. Then there exists $x_0 \in [0,1]$ such that $z(x_0) \in (\nu_1, \nu_2)$. Due to the fact that z satisfies the first and second Weierstrass–Erdmann corner–conditions, z has to be continuous at $x = x_0$. This implies the existence of an interval $I \subseteq [0,1]$ with $x_0 \in I$ and $z_{|I} \in (\nu_1, \nu_2)$.
Let φ be a test function with $\operatorname{supp}\varphi \subseteq I$ and therefore $\psi(x) := \int_0^x \varphi(s)ds$ is a suitable test function for (P_ε). With respect to the assumed stability of u_n we have

$$0 \leq \lim_{n\to\infty} D^2 J(u_n)(\psi,\psi) = \lim_{n\to\infty} \int_0^1 \varepsilon_n \varphi'^2 + \sigma'(z_n)\varphi^2 dx = \int_I \sigma'(z)\varphi^2 dx < 0, \quad (4.23)$$

a contradiction. Note, that the limit in (4.23) was achieved by Lebesgue's dominated convergence theorem.
ad (ii): This statement is an easy consequence of Lemma 62. $\qquad\square$

Now we state another hypothesis on b:

(H–2) Let $X = Y$, where X is defined by (4.9) and Y by (4.11). Moreover we assume the following for $A_j = [x_1^j, x_2^j]$, $j = 1, \ldots, r$: Either

$$\int_{x_1^j}^1 b(\tau)d\tau + t_r < \sigma(\nu_a) \text{ and } \int_{x_2^j}^1 b(\tau)d\tau + t_r > \sigma(\nu_b) \qquad (4.24)$$

or

$$\int_{x_1^j}^1 b(\tau)d\tau + t_r > \sigma(\nu_b) \text{ and } \int_{x_2^j}^1 b(\tau)d\tau + t_r < \sigma(\nu_a) \qquad (4.25)$$

holds, where ν_a and ν_b are defined by (4.7).
Let $\hat{\Omega} := [0,1]\backslash\bigcup_{j=1}^r A_j$ and let $\Omega_l \subseteq \hat{\Omega}$ a nodal domain of b. Then we assume either

$$\int_x^1 b(\tau)d\tau + t_r \geq \sigma(\nu_b) \qquad (4.26)$$

or

$$\int_x^1 b(\tau)d\tau + t_r \leq \sigma(\nu_a) \qquad (4.27)$$

for every $x \in \Omega_l$.

Note, that (H–2) is a much stronger assumption than (H–1). In particular, if (H–2) is satisfied (H–1) is also fulfilled.

Lemma 60. *We assume that b satisfies (H–2). Then there exists a unique solution $v \in L^\infty(0,1)$ of the equation*

$$\int_x^1 b(\tau)d\tau + t_r = \sigma(v(x)) \qquad (4.28)$$

provided v satisfies the first and the second Weierstrass–Erdmann corner conditions.

Proof. Let v be a solution of (4.28). Hypothesis (H–2) and the monotonicity of $x \mapsto \int_x^1 b(\tau)d\tau + t_r$ in A_j yields that v has to jump in each A_j. Furthermore v satisfies the corner–conditions and therefore

$$v \text{ jumps from } \nu_m \text{ to } \nu_M \text{ or vice versa,}$$

v jumps in each A_j exactly once and v does not jump in $[0,1]\backslash\bigcup_{j=1}^r A_j$ by (4.26) and (4.27). This determines v uniquely and the proof is done. $\square$

This observation yields by virtue of Lemma 58 at once the following:

Corollary 61. *Let b satisfy (H–2) and consider any sequence $(z_n)_{n\in\mathbb{N}}$ of stable solutions of* (4.17). *Then $(z_n)_{n\in\mathbb{N}}$ converges pointwise to the global minimizer of* (P).

The last result in this section considers convergence in energy and its convergence–rate with respect to ε. The proof follows the lines of the proof of Lemma 2 in [4] using a suitable comparison function.

Lemma 62. *Let $\tilde{u}_\varepsilon$ be a global minimizer of (P_ε) and let u be the global minimizer of* (P), *then the following inequalities are valid with a constant $C > 0$ independent of ε:*

(i)

$$0 \le \int_0^1 \left(\frac{\varepsilon}{2}\tilde{u}_\varepsilon''^2 + W(\tilde{u}_\varepsilon') - b\tilde{u}_\varepsilon - W(u') + bu \right) dx - t_r\tilde{u}_\varepsilon(1) + t_r u(1) \le C\sqrt{\varepsilon} \quad (4.29)$$

(ii)

$$0 \le \int_0^1 \left(W(\tilde{u}_\varepsilon') - b\tilde{u}_\varepsilon - W(u') + bu \right) dx - t_r\tilde{u}_\varepsilon(1) + t_r u(1) \le C\sqrt{\varepsilon} \quad (4.30)$$

(iii)

$$0 \le \int_0^1 \frac{\varepsilon}{2}\tilde{u}_\varepsilon''^2 \le C\sqrt{\varepsilon} \quad (4.31)$$

Proof. We seperate the proof into two steps.

Step 1: (Construction of a suitable comparison function u_ε):
Let u be the global minimizer of (P) and let $x_1,\ldots,x_r$ be the jump points of $z = u'$. We define for every $j = 1,\ldots,r$ the sets $A_j^\varepsilon := [x_j - \sqrt{\varepsilon}, x_j + \sqrt{\varepsilon}]$ and $A^\varepsilon := \bigcup_{j=1}^r A_j^\varepsilon$. Furthermore we define the following function

$$u_\varepsilon'(x) := \begin{cases} u'(x), & \text{if } x \in [0,1]\backslash A^\varepsilon \\ \frac{1}{2\sqrt{\varepsilon}}\big[-(x - (x_j + \sqrt{\varepsilon}))\, u'(x_j - \sqrt{\varepsilon}) + (x - (x_j - \sqrt{\varepsilon}))\, u'(x_j + \sqrt{\varepsilon})\big], \\ \qquad\qquad \text{if } x \in A_j^\varepsilon. \end{cases} \quad (4.32)$$

Now we define $u_\varepsilon(x) := \int_0^x u_\varepsilon'(s)ds$ and

$$u_\varepsilon \in \mathcal{A} = H^2(0,1) \cap \{v \mid v(0) = 0\}$$

is obvious. The second derivative reads as follows:

$$u_\varepsilon''(x) := \begin{cases} u''(x) & , \text{ if } x \in [0,1]\backslash A^\varepsilon \\ \frac{1}{2\sqrt{\varepsilon}}\big[-u'(x_j - \sqrt{\varepsilon}) + u'(x_j + \sqrt{\varepsilon})\big] & , \text{ if } x \in A_j^\varepsilon. \end{cases} \quad (4.33)$$

Step 2:

(i) First we show the estimate $\int_0^1 \frac{\varepsilon}{2} u_\varepsilon''^2 dx \leq C\sqrt{\varepsilon}$:
   Trivially we have

$$\int_0^1 \frac{\varepsilon}{2} u_\varepsilon''^2 dx = \int_{[0,1]\backslash A_\varepsilon} \frac{\varepsilon}{2} u_\varepsilon''^2 dx + \int_{A_\varepsilon} \frac{\varepsilon}{2} u_\varepsilon''^2 dx =: J_1 + J_2.$$

For $x \in [0,1]\backslash A_\varepsilon$ we obtain the bound $||u_\varepsilon''(x)|| \leq K$ with K independent of ε and this yields $J_1 \leq K\varepsilon$. Also for J_2 we get the following bound by considering (4.33)

$$J_2 = \sum_{j=1}^r \int_{x_j-\sqrt{\varepsilon}}^{x_j+\sqrt{\varepsilon}} \varepsilon \frac{1}{4\varepsilon} \left(-u'(x_j - \sqrt{\varepsilon}) + u'(x_j + \sqrt{\varepsilon}) \right)^2 dx = \frac{r}{2}\sqrt{\varepsilon} \, ||u_\varepsilon'||_\infty \leq K\sqrt{\varepsilon}.$$

(ii) We claim the inequality $\left|\left| \int_0^1 \left(W(u_\varepsilon') - t_r u_\varepsilon' - W(u') + t_r u' \right) dx \right|\right| \leq C\sqrt{\varepsilon}$.
   Because of $u_\varepsilon'(x) = u'(x)$ for every $x \in [0,1]\backslash A_\varepsilon$ we obtain by the mean value theorem and the uniform bound for u_ε' and u' the estimate

$$\left|\left| \int_0^1 \left(W(u_\varepsilon') - t_r u_\varepsilon' - W(u') + t_r u' \right) dx \right|\right| =$$

$$= \left|\left| \int_{A_\varepsilon} \left(W(u_\varepsilon') - t_r u_\varepsilon' - W(u') + t_r u' \right) dx \right|\right| \leq Cr \left| A_j^\varepsilon \right| \leq C\sqrt{\varepsilon}.$$

(iii) It remains to show $\left|\left| \int_0^1 b(x)(u_\varepsilon - u) dx \right|\right| \leq C\sqrt{\varepsilon}$.
   To prove this inequality we use the fundamental theorem of calculus:

$$\left|\left| \int_0^1 b(x)(u_\varepsilon - u) dx \right|\right| \leq ||b||_\infty \int_0^1 \left|\left| \int_0^x (u_\varepsilon' - u') ds \right|\right| dx =$$

$$||b||_\infty \int_0^1 \left|\left| \int_{[0,x]\cap A_\varepsilon} (u_\varepsilon' - u') ds \right|\right| dx \leq K \left| A_\varepsilon \right| \leq K\sqrt{\varepsilon}.$$

Inequalities (i)-(iii) yields (4.29). Furthermore (4.30) can be easily established by the trivial fact $\int_0^1 \frac{\varepsilon}{2} u_\varepsilon''^2 dx \geq 0$. Note, that u is the global minimizer of (P) and this yields (4.31).   □

4.4   Shape of stable critical points

In this section we discuss the shape of stable solutions z_ε of (4.17) for ε sufficiently small. In particular we determine the exact number of zeros of z_ε' and moreover we give a closer description of the intervals in which z_ε crosses $[\nu_m, \nu_M]$.

Let $\{y_1, \ldots, y_s\}$ the zeros of b in $(0,1)$. This implies the existence of $\Omega_1, \ldots, \Omega_{s+1}$ such that $\text{sign}\, b_{|\Omega_j} \equiv const.$ and we denote $\Omega_j := (y_{j-1}, y_j]$ for $j = 1, \ldots, s+1$ with $y_0 = 0$ and $y_{s+1} = 1$.

In the sequel let z be a stable solution of (4.17).

The first lemma is crucial in our forthcoming analysis, but its proof is unfortunately rather technical.

Lemma 63. *We assume, that z' changes sign at most $s-1$ times. Then there exists Ω_j such that* $\operatorname{sign}(z'b)_{|\Omega_j} > 0$.

Proof. We assume, that z' changes sign exactly $s-1-k$ times for a certain $k \geq 1$ and we denote these intervals in which a change in sign takes place by $\Omega_{j_1}, \ldots, \Omega_{j_t}$ with $t \leq s - 1 - k$. Furthermore we define for $l = 1, \ldots, t$ the set

$$\Omega_{j_l} := (y_{j_l - 1}, y_{j_l}]$$

and without loss of generality we assume $y_{j_l} < y_{j_m}$ for $l < m$. By our assumptions we know the existence of $s + 1 - t$ nodal domains of b in which no sign change of z' will occur and we denote them by $\Omega_{n(1)}, \ldots, \Omega_{n(s+1-t)}$. Let

$$\Omega_{n(k)} := (y_{n(k)-1}, y_{n(k)}]$$

such that $y_{n(k)} < y_{n(l)}$ for $k < l$.

We consider $\Omega_{n(k)}, \Omega_{n(k+1)}$ for $k \in \{1, \ldots, s-t\}$ and the convex hull of the two sets can be denoted by the above definitions as $(y_{n(k)-1}, y_{n(k+1)}]$. Note, that if $y_{n(k+1)} = y_{n(k)+1}$ is valid, the statement in Lemma 63 is trivial, because we obtain two connected nodal domains of b in which z' does not change sign. Thus we can assume without loss of generality $y_{n(k+1)} > y_{n(k)+1}$ and this implies the existence of $\Omega_{j_l}, \ldots, \Omega_{j_m}$ such that the following representation holds:

$$(y_{n(k)-1}, y_{n(k+1)}] = \bigcup_{r=l}^{m} \Omega_{j_r} \cup \Omega_{n(k)} \cup \Omega_{n(k+1)}.$$

We claim the existence of a certain $h \in \{1, \ldots, s-t\}$ such that z' has exactly one change in sign in Ω_{j_r} for $r = l(h), \ldots, m(h)$ and where

$$(y_{n(h)-1}, y_{n(h+1)}] = \bigcup_{r=l(h)}^{m(h)} \Omega_{j_r} \cup \Omega_{n(h)} \cup \Omega_{n(h+1)}$$

holds. Suppose this is not the case. Then we have at least

$$m(h) - l(h) + 2 = n(h+1) - n(h)$$

changes in sign between $\Omega_{n(h)}$ and $\Omega_{n(h+1)}$. This yields the number of sign changes of z' is bigger or equal to

$$n(1) - 1 + n(2) - n(1) + \cdots + n(s+1-t) - n(s-t) + s + 1 - n(s+1-t) = s$$

contradicting our assumption.

Hence there exists $\Omega_{n(h)} = \Omega_p$ and $\Omega_{n(h+1)} = \Omega_r$ with $r = p + v$ for a certain $v \in \mathbb{N}$, such that z' changes sign exactly one time in $\Omega_{p+1}, \ldots, \Omega_{p+v-1}$ and this implies

$$\begin{aligned}
\operatorname{sign} b_{|\Omega_{p+v}} &= (-1)^v \operatorname{sign} b_{|\Omega_p}, \\
\operatorname{sign} z'_{|\Omega_{p+v}} &= (-1)^{v-1} \operatorname{sign} z'_{|\Omega_p}.
\end{aligned} \tag{4.34}$$

Hence one obtains

$$\operatorname{sign} bz'_{|\Omega_p} > 0 \quad \text{or} \quad \operatorname{sign} bz'_{|\Omega_r} > 0$$

and the proof is done. $\qquad \square$

By Lemma 63 we can prove the following:

Lemma 64. *Let z_ε be a stable solution of* (4.17). *Then there exists $\varepsilon_0 > 0$ sufficiently small, such that z'_ε has exactly s changes in sign for every $\varepsilon < \varepsilon_0$ in the interval $(0, 1)$.*

Proof. We prove by contradiction and thus we assume the existence of a sequence $(z_n)_{n \in \mathbb{N}}$ for $\varepsilon_n \searrow 0$ and z'_n has at most $s - 1$ changes of sign for every $n \in \mathbb{N}$. (Note, that it is impossible for z'_n to have more than s sign changes by Corollary 57.) By virtue of Lemma 63 we obtain for each $n \in \mathbb{N}$ the existence of $\Omega_{j_{\varepsilon_n}}$, a nodal domain of b, such that

$$\text{sign}(bz'_n)_{|\Omega_{j_{\varepsilon_n}}} > 0$$

holds. Because the number of the Ω_j is finite, there exists a subsequence, denoted without loss of generality by $(z_n)_{n \in \mathbb{N}}$, and a certain Ω_l for $l \in \{1, \ldots, s + 1\}$ such that

$$\text{sign}(z'_n b)_{|\Omega_l} > 0 \tag{4.35}$$

for every $n \in \mathbb{N}$. Without loss of generality we assume $b_{|\Omega_l} > 0$ and (4.35) implies $z'_{n|\Omega_l} > 0$. Furthermore by virtue of Lemma 58 we have $\lim_{n \to \infty} z_n(x) = z(x)$ for some $z \in L^\infty(0, 1)$ and for every $x \in [0, 1]$ and one obtains

$$0 = \lim_{n \to \infty} \left(\varepsilon_n z''_n - \sigma(z_n) + \int_x^1 b(\tau) d\tau + t_r \right) = -\sigma(z) + \int_x^1 b(\tau) d\tau + t_r. \tag{4.36}$$

Because $b_{|\Omega_l} > 0$ we get $x \mapsto \int_x^1 b(\tau) d\tau + t_r$ is strictly monotonically decreasing in Ω_l and (4.36) yields

$$\sigma(z)_{|\Omega_l} \text{ is also strictly monotonically decreasing.} \tag{4.37}$$

On the other hand we have $z'_{n|\Omega_l} > 0$ for every $n \in \mathbb{N}$ and Helly's theorem implies z is monotonically increasing in Ω_l. With respect to the shape of σ it follows immediately

$$\text{either } z_{|\Omega_l} \in (\nu_1, \nu_2) \text{ or } z_{|\Omega_l} \equiv c.$$

The first can be excluded by Lemma 59 and the second one by (4.37). $\square$

Remark 65. Note, that by the above proof the existence of a nodal domain Ω_l of b and a sequence $(z_n)_{n \in \mathbb{N}}$ of stable stable solutions of (4.17) such that $\text{sign}(z'_n b)_{|\Omega_l} > 0$ for every $n \in \mathbb{N}$ is impossible.

Next we discuss the zeros of z'_ε. Recall, that b has exactly s zeros.

Theorem 66. *Let z_ε be a stable critical point of (P_ε). Then there exists ε_0 sufficiently small, such that for every $\varepsilon < \varepsilon_0$ we obtain the existence of exactly s non–degenerate zeros of z'_ε in $(0, 1)$.*

Proof. By Lemma 64 we just have to show that every zero of z'_ε is non–degenerate. Let z_ε be any stable solution of (4.17) with ε sufficiently small. For the rest of the proof we suppress ε for our convenience. We have to distinguish two cases:

(i) We assume the existence of $x_0 \in (0,1)$ such that

$$z'(x_0) = 0 \text{ and } z' \text{ does not change sign at } x_0,$$

which implies $z''(x_0) = 0$. Without loss of generality we assume that $z'(x) \geq 0$ for every $x \in (x_0 - \delta, x_0 + \delta)$ for some δ sufficiently small. Again we have to distinguish several cases:

(a) Let $b(x_0) < 0$:
By differentiating (4.17) we obtain that z' satisfies the following equation in $(x_0 - \delta, x_0 + \delta)$:

$$\varepsilon(z')'' - \sigma'(z)z' = b < 0,$$
$$z'_{|(x_0 - \delta, x_0 + \delta)} \geq 0, \quad z''(x_0) = 0. \tag{4.38}$$

We deduce by Theorem 34 $z' \equiv 0$ in $(x_0 - \delta, x_0 + \delta)$ which is impossible due to (4.38) and $b_{|(x_0 - \delta, x_0 + \delta)} < 0$.

(b) $b(x_0) = 0$:
This assumption implies that b changes sign at $x = x_0$. By the same procedure like in (a) we obtain $z' \equiv 0$ either in $(x_0 - \delta, x_0)$ or in $(x_0, x_0 + \delta)$, again a contradiction.

(c) $b(x_0) > 0$:
Let Ω_{x_0} be the nodal domain of b containing x_0. Because of $z'(x) \geq 0$ for every $x \in (x_0 - \delta, x_0 + \delta)$ and the stability of z we obtain by Lemma 55 $z'_{|\Omega_{x_0}} \geq 0$. Hence we have $\text{sign}(z'b)_{|\Omega_{x_0}} \geq 0$ which is impossible for ε sufficiently small by Remark 65.

(ii) We assume the existence of $x_0 \in (0,1)$ with

$$z'(x_0) = z''(x_0) = 0 \text{ and } z' \text{ changes sign at } x_0$$

(without loss of generality from $+$ to $-$).

(a) We treat the cases $b(x_0) > 0$ and $b(x_0) < 0$ in an analogues manner:
Again (4.38) and Hopf's maximum principle yields $z' \equiv 0$ either in $(x_0 - \delta, x_0)$ or in $(x_0, x_0 + \delta)$ for δ sufficiently small. This is again a contradiction.

(b) Let $b(x_0) = 0$:
Here we have to distinguish two cases: The first one deals with b changes sign at x_0 from $-$ to $+$. Then we get a contradiction again by (4.38) just by applying the maximum principle.
In the second case we suppose b changes sign from $+$ to $-$ like z'. By the stability assumption this implies the existence of a nodal domain Ω_l of b such that $\text{sign}(z'b)_{|\Omega_l} > 0$ and this is again a contradiction for ε small by virtue of Remark 65.

$\square$

The next goal is to study the behavior of z_ε, if the function crosses the interval $[\nu_m, \nu_M]$. First we state an easy consequence of (H-1): Due to continuity and monotonicity of b, for every $A_j = [x_1^j, x_2^j]$ there exists $\delta_j > 0$ such that either

$$\int_{x_1^j}^1 b(\tau)d\tau + t_r < \sigma(\nu_m) - \delta_j \quad \text{and} \quad \int_{x_2^j}^1 b(\tau)d\tau + t_r > \sigma(\nu_m) + \delta_j \qquad (4.39)$$

or

$$\int_{x_1^j}^1 b(\tau)d\tau + t_r > \sigma(\nu_m) + \delta_j \quad \text{and} \quad \int_{x_2^j}^1 b(\tau)d\tau + t_r < \sigma(\nu_m) - \delta_j \qquad (4.40)$$

holds.

Definition 67. We say a sequence $(z_n)_{n \in \mathbb{N}}$ fulfills property (E), if one of the following possibilities are satisfied:

(i) Let z_n be a global minimizer of (P_{ε_n}) for every $n \in \mathbb{N}$ and b satisfies (H–1).

(ii) For every $n \in \mathbb{N}$ let z_n be a stable solution of (4.17) and furthermore let (H–2) be fulfilled.

Remark 68. In particular we obtain by virtue of Lemma 59 and Corollary 61, that every sequence $(z_n)_{n \in \mathbb{N}}$ satisfying property (E) converges to the global minimizer of (P).

Lemma 69. Let $(z_\varepsilon)_\varepsilon$ satisfy property (E). Then we deduce for $\varepsilon < \varepsilon_0$ with ε_0 sufficiently small that z_ε crosses $[\nu_m, \nu_M]$ in the interior of $A_j = [x_1^j, x_2^j]$. In particular, we have either

$$z_\varepsilon(x_1^j) < \nu_m \quad \text{and} \quad z_\varepsilon(x_2^j) > \nu_M, \qquad (4.41)$$

if $x \mapsto \int_x^1 b(\tau)d\tau + t_r$ is strictly monotonically increasing in A_j or

$$z_\varepsilon(x_1^j) > \nu_M \quad \text{and} \quad z_\varepsilon(x_2^j) < \nu_m, \qquad (4.42)$$

if $x \mapsto \int_x^1 b(\tau)d\tau + t_r$ is strictly monotonically decreasing in A_j.

Proof. Without loss of generality we assume $b_{|A_j} > 0$ and this implies $x \mapsto \int_x^1 b(\tau)d\tau + t_r$ is strictly monotonically decreasing in A_j. Due to (H–1) we have

$$\int_{x_1^j}^1 b(\tau)d\tau + t_r > \sigma(\nu_m) + \delta \quad \text{and} \quad \int_{x_2^j}^1 b(\tau)d\tau + t_r < \sigma(\nu_m) - \delta \qquad (4.43)$$

for a certain $\delta > 0$. Obviously it is enough to show (4.42) for $\varepsilon \le \varepsilon_0$ and we suppose this is not the case. Let $(z_n)_{n \in \mathbb{N}}$ be a sequence with property (E) and assume (4.42) is not fulfilled.

First we consider the case $z_n(x_2^j) \ge \nu_m$ for every $n \in \mathbb{N}$. Because of

$$z_n(x_2^j) \to z(x_2^j) \ge \nu_m,$$

where z is the global minimizer of (P), we obtain either $z(x_2^j) = \nu_m$ or $z(x_2^j) \ge \nu_M$. Both cases imply (remember the shape of σ and (4.43))

$$\sigma(\nu_m) \le \sigma(z(x_2^j)) = \int_{x_2^j}^1 b(\tau)d\tau + t_r < \sigma(\nu_m) - \delta,$$

a contradiction to $\delta > 0$.

The other case can be treated in an analogues manner. $\qquad\qquad\qquad\qquad\qquad \square$

Remark 70. Let

$$\hat{\Omega} := [0,1]\backslash \bigcup_{j=1}^{r} A_j, \tag{4.44}$$

where A_j is defined by (4.10) and furthermore let b satisfy either (H–1) or (H–2). Then it is easy to see that

$$\int_x^1 b(\tau)d\tau + t_r \neq \sigma(\nu_m) \quad \text{holds for every } x \in \hat{\Omega}. \tag{4.45}$$

In particular, $z \in C^2(\hat{\Omega})$ by (4.14), if z denotes the global minimizer of (P).

Lemma 71. *Let z_ε be a stable solution of (4.17) and moreover let b satisfy property (H–2). Then for every $x \in \hat{\Omega}$ and for $\varepsilon \leq \varepsilon_0$, ε_0 sufficiently small, we obtain $z_\varepsilon(x) \notin [\nu_m, \nu_M]$.*

Proof. We prove by contradiction:
Let $\Omega_j := [\alpha_j, \beta_j]$ be a nodal domain of b with $\Omega_j \subseteq \hat{\Omega}$ and without loss of generality we assume for every $x \in \Omega_j$

$$\int_x^1 b(\tau)d\tau + t_r > \sigma(\nu_b), \tag{4.46}$$

because b satisfies (H–2) by assumption. Obviously it is enough to show the assertion for Ω_j. Therefore we assume the existence of a sequence $(z_n)_{n\in\mathbb{N}}$ and $x_n \in \Omega_j$ such that $z_n(x_n) \in [\nu_m, \nu_M]$ for every $n \in \mathbb{N}$. Due to the pointwise convergence of $(z_n)_{n\in\mathbb{N}}$ to the global minimizer z of (P) we have for $\vartheta \in \{\alpha_j, \beta_j\}$ and for $n > n_0$

$$z_n(\vartheta) > \nu_b$$

by the shape of σ and (4.46). By our assumption we obtain $z_{n|\Omega_j}$ has a global minimum at $\hat{x}_n$ with $z_n(\hat{x}_n) \leq \nu_M$ and $z_n''(\hat{x}_n) \geq 0$. The Euler–Lagrange equation yields

$$\sigma(\nu_b) < \int_{\hat{x}_n}^1 b(\tau)d\tau + t_r =$$
$$= -\varepsilon_n z_n''(\hat{x}_n) + \sigma(z_n(\hat{x}_n)) \leq \sigma(z_n(\hat{x}_n)) \leq \sigma(\nu_1) = \sigma(\nu_b).$$

Thus the Lemma is proved. □

Before continuing the discussion of the qualitative behavior of stable solutions of (P_ε) we have to study the unperturbed variational problem (P) which reads as follows

$$J(u) := \int_0^1 W(u') - b(x)u\, dx - t_r u(1),$$
$$u(0) = 0.$$

We mention, that the forthcoming results will be very useful in the next section, too. In the sequel let $z = u'$ and by integration by parts and by the fundamental theorem in calculus we rewrite (P) in an equivalent form

$$\tilde{J}(z) := \int_0^1 \left(W(z) - \int_x^1 b(\tau)d\tau\, z - t_r z \right) dx. \tag{4.47}$$

The global minimizer or a stable critical point of (P) is given by $u(x) = \int_0^x z(s)ds$ since the relation $J(u) = \tilde{J}(u')$ is valid for every $u \in W^{1,p}(0,1) \cap \{v \mid v(0) = 0\}$. Furthermore we define

$$(x,z) \mapsto \hat{W}(x,z) := W(z) - \int_x^1 b(\tau)d\tau z - t_r z$$

$$\hat{W} : [0,1] \times \mathbb{R} \mapsto \mathbb{R}$$

(4.48)

and we obtain

$$\tilde{J}(z) = \int_0^1 \hat{W}(x, z(x))dx = J(u)$$

(4.49)

with $z \in L^p(0,1)$ and $u(x) = \int_0^x z(s)ds$.

Our goal is to show, that the global minimizer of (P) minimizes also the integrand pointwise. To achieve that it is convenient to rewrite $\hat{W}$ as follows: We define

$$\vartheta = \vartheta(x) := \int_x^1 b(\tau)d\tau + t_r$$

and we obtain

$$\tilde{W}(\vartheta, z) := W(z) - \vartheta z.$$

(4.50)

The properties of (4.50) are discussed in [9], Proposition 2.1, and this is the statement:

Proposition 72. *The function $\tilde{W} = \tilde{W}(\vartheta, z)$ is as regular as W and ϑ. Furthermore we claim for $\vartheta \in (\sigma(\nu_a), \sigma(\nu_b))$:*

(i) *There exists exactly three critical points $\alpha_\vartheta < \beta_\vartheta < \gamma_\vartheta$ such that $\frac{\partial}{\partial z}\tilde{W}(\vartheta, \delta_\vartheta) = 0$ for $\delta = \alpha, \beta, \gamma$. Furthermore we have $\alpha_\vartheta \in (\nu_a, \nu_1)$ and $\gamma_\vartheta \in (\nu_2, \nu_b)$ are local minima, $\beta_\vartheta \in (\nu_1, \nu_2)$ is a local maximum.*

(ii) *$\tilde{W}(\vartheta, \cdot)$ is strictly monotonically increasing in $(\alpha_\vartheta, \beta_\vartheta) \cup (\gamma_\vartheta, \infty)$ and strictly monotonically decreasing elsewhere.*

(iii) *The identity $\vartheta_0 = \sigma(\nu_m)$ implies*

$$\tilde{W}(\vartheta_0, \nu_m) = \tilde{W}(\vartheta_0, \nu_M).$$

(4.51)

For $\vartheta < \vartheta_0$ we have

$$\tilde{W}(\vartheta, \alpha_\vartheta) < \tilde{W}(\vartheta, \gamma_\vartheta)$$

(4.52)

and in particular $\alpha_\vartheta \in (\nu_a, \nu_m)$. On the other hand $\vartheta > \vartheta_0$ yields

$$\tilde{W}(\vartheta, \alpha_\vartheta) > \tilde{W}(\vartheta, \gamma_\vartheta)$$

(4.53)

with $\gamma_\vartheta \in (\nu_m, \nu_b)$.

For the proof we refer to [9].

The typical shape of $\tilde{W}$ depending on ϑ can be seen in the following picture:

$\vartheta < \vartheta_0$ $\vartheta = \vartheta_0$ $\vartheta > \vartheta_0$

Proposition 73. *(i) For $\vartheta > \sigma(\nu_b)$ respectively $\vartheta < \sigma(\nu_a)$ the function $\tilde{W}(\vartheta, \cdot)$ possesses exactly one global minimum at $z = (\sigma)^{-1}(\vartheta)$.*

(ii) If $\vartheta \in \{\sigma(\nu_a), \sigma(\nu_b)\}$, then $\tilde{W}(\vartheta, \cdot)$ has also exactly one global minimum at $z = \nu_a$ respectively $z = \nu_b$.

We omit the proof because it is analogues to the proof of Proposition 72.

Corollary 74. *The global minimizer z of (P) jumps from ν_m to ν_M at $x = x_0$ or vice versa, if the identity $\hat{W}(x_0, \nu_m) = \hat{W}(x_0, \nu_M)$ holds.*

Proof. By the assumptions on b we obtain that the minimizer z jumps at $x = x_0$ if and only if $\int_{x_0}^{1} b(\tau)d\tau + t_r = \sigma(\nu_m)$ holds. Furthermore by Proposition 72(iii) this equation is just valid for $\hat{W}(x_0, \nu_m) = \hat{W}(x_0, \nu_M)$. $\qquad\square$

The next statement shows that the minimizer of (P) also minimizes the integrand pointwise.

Corollary 75. *Let z be the global minimizer of (P). Then for an arbitrary $x \in [0,1]$ the following inequality is valid for every $v \in \mathbb{R}$:*

$$\hat{W}(x, z(x)) \leq \hat{W}(x, v). \tag{4.54}$$

Proof. Let z be the global minimizer of (P) and thus z satisfies

$$\frac{\partial}{\partial z}\hat{W}(x, z(x)) = 0$$

for every $x \in [0,1]$. Furthermore we have (see (4.14))

$$z(x) \begin{cases} < \nu_m, & \text{if } \int_x^1 b(\tau)d\tau + t_r < \sigma(\nu_m) \\ > \nu_M, & \text{if } \int_x^1 b(\tau)d\tau + t_r > \sigma(\nu_m). \end{cases}$$

Thus Propositions 72 and 73 yield the result. $\qquad\square$

Moreover we need the following:

Lemma 76. *Let z be the global minimizer of (P) and let $\Omega \subseteq [0,1]$ be a nodal domain of b such that $z_{|\Omega} > \nu_M$ (in particular, no jump of z takes place in Ω). Furthermore we define*

$$J_{x,\gamma} := (z(x) - \gamma, z(x)).$$

Then there exists $\gamma > 0$, such that for every $x \in \Omega$ and for all $r \in (-\infty, z(x) - \gamma)$ and $q \in J_{x,\gamma}$ we have

$$\hat{W}(x, r) \geq \hat{W}(x, q), \tag{4.55}$$

where $\hat{W}$ is defined as in (4.48).
An similar statement holds for $z_{|\Omega} < \nu_m$.

Proof. We assume the above assertion is wrong. Then we obtain for every $n \in \mathbb{N}$ by setting $\gamma = \frac{1}{n}$ a certain $x_n \in \Omega$, $r_n \in (-\infty, z(x_n) - \frac{1}{n})$ and $q_n \in J_{x_n, \gamma_n}$ with

$$\hat{W}(x_n, r_n) < \hat{W}(x_n, q_n). \tag{4.56}$$

Because $(x_n)_{n \in \mathbb{N}}$ is a bounded sequence we get (for a subsequence) $x_n \to x_0 \in \bar{\Omega}$ and due to $z_{|\Omega}$ is continuous we obtain $z(x_n) \to z(x_0)$. This implies immediately

$$\lim_{n \to \infty} q_n = z(x_0)$$

and moreover

$$\lim_{n \to \infty} \hat{W}(x_n, q_n) = \hat{W}(x_0, z(x_0)).$$

By virtue of (4.56) we obtain that the sequence $(r_n)_{n \in \mathbb{N}}$ is bounded and therefore there exists $r_0 \in (-\infty, z(x_0))$ with $r_n \to r_0$. Because of

$$\lim_{n \to \infty} \hat{W}(x_n, r_n) = \hat{W}(x_0, r_0) \le \hat{W}(x_0, z(x_0)),$$

we have $r_0 = z(x_0)$ by Corollary 75.

Hence, there exists for arbitrary $\delta > 0$ a certain $n_0(\delta)$ such hat for every $n > n_0$ we have

$$r_n \in \left(z(x_0) - \delta, z(x_n) - \frac{1}{n} \right).$$

We claim for every $r_n \in (z(x_0) - \delta, z(x_n) - \frac{1}{n})$ and $q_n \in (z(x_n) - \frac{1}{n}, z(x_n))$ the estimate

$$\hat{W}(x_n, r_n) > \hat{W}(x_n, q_n). \tag{4.57}$$

First note, that there exists $c(n) > \frac{1}{n}$ sufficiently small with $z(x_0) - \delta = z(x_n) - c(n)$. Furthermore we have

$$\frac{\partial}{\partial z} \hat{W}(x_n, z)_{|(z(x_n) - c(n), z(x_n))} = \left(W'(z) - \int_{x_n}^{1} b(\tau) d\tau - t_r \right)_{|(z(x_n) - c(n), z(x_n))} =$$
$$= \left(W'(z) - W'(z(x_n)) \right)_{|(z(x_n) - c(n), z(x_n))} < 0$$

due to the shape of W'. This implies (4.57), thus a contradiction to (4.56). $\qquad\square$

With this technical Lemma at hand we can prove the following:

Lemma 77. *Let ε_0 be sufficiently small, let $(z_\varepsilon)_\varepsilon$ be a sequence of global minimizers of (P_ε), let z be the global minimizer of (P) and let b satisfy (H–1). Furthermore let $\Omega \subseteq [0,1]$ be a nodal domain of b where no jump of z occurs. We define*

$$\nu_r := \sigma^{-1}(\nu_M)$$

with $\nu_r \in (\nu_1, \nu_2)$ and

$$\Gamma := \min \{ |\nu_M - \nu_r|, |\nu_r - \nu_m| \}.$$

Furthermore we assume

$$\frac{||b||_{\infty, \Omega}}{\min \{ \sigma'(\nu_m), \sigma'(\nu_M) \}} < \sqrt{2} \frac{\Gamma}{|\Omega|}, \tag{4.58}$$

then $z_\varepsilon(x) \notin (\nu_m, \nu_M)$ for every $\varepsilon < \varepsilon_0$ and for every $x \in \Omega$.

Proof. We assume the existence of a sequence $(z_n)_{n\in\mathbb{N}}$ of global minimizers of (P_{ε_n}) and $x_n \in \Omega$ with $z_n(x_n) \in (\nu_m, \nu_M)$ for every $n \in \mathbb{N}$. Without loss of generality we assume the global minimizer z of (P) strictly monotonically decreasing in Ω and

$$\int_x^1 b(\tau)d\tau + t_r > \sigma(\nu_M) + \delta$$

for fixed $\delta > 0$ and for every $x \in \Omega := [\alpha, \beta]$.

Due to the pointwise convergence $z_n \to z$, where z denotes the global minimizer, we obtain

$$z_n(\alpha), z_n(\beta) \geq \nu_M \text{ for every } n \in \mathbb{N}$$

and because of our assumption we deduce a minimum (without loss of generality at x_n) such that

$$z_n(x_n) < \nu_M \text{ and } z_n''(x_n) \geq 0.$$

The Euler–Lagrange equation yields

$$\sigma(z_n(x_n)) = \sigma(z(x_n)) + \varepsilon_n z_n''(x_n) \geq \sigma(z(x_n)) > \sigma(\nu_M)$$

and therefore we have due to the shape of σ for every $n \in \mathbb{N}$

$$z_n(x_n) \in (\nu_m, \nu_r). \tag{4.59}$$

By virtue of Lemma 76 there exists $\gamma > 0$ such that

$$\hat{W}(x, \tilde{z}) \geq \hat{W}(x, \hat{z}) \tag{4.60}$$

holds for every $x \in \Omega$, $\hat{z} \in (z(x) - \gamma, z(x))$ and $\tilde{z} \in (-\infty, z(x) - \gamma)$. Due to the pointwise convergence we have

$$||z_n(\omega) - z(\omega)|| \leq \frac{\gamma}{2}$$

for $\omega = \{\alpha, \beta\}$. By our assumption we get existence of $\xi_1, \xi_2 \in \Omega$ such that $x_n \in (\xi_1, \xi_2)$ and

$$z_n(\xi_j) - z(\xi_j) = -\gamma,$$
$$(z_n - z)_{|(\xi_1, \xi_2)} < -\gamma$$

for $j = 1, 2$. Furthermore we define $M := ||z'||_{\infty, \Omega}$ and by differentiating (4.21) we obtain

$$M \leq \frac{||b||_\infty}{\min\{\sigma'(\nu_m), \sigma'(\nu_M)\}}. \tag{4.61}$$

Our next aim is to construct a competitor $\tilde{z}$ of z_n:

We impose a partition of (ξ_1, ξ_2) in disjoint intervals such that

$$(\xi_1, \xi_2) = \bigcup_{k=1}^m [x_k, y_{k+1}] \cup \bigcup_{k=1}^m [y_k, x_k]$$

with $y_1 = \xi_1$ and $y_{m+1} = \xi_2$. Furthermore we define

$$\tilde{z}'(\xi) := \begin{cases} 0, & \text{if } x \in \bigcup_{k=1}^m (y_k, x_k) \\ -M, & \text{elsewhere,} \end{cases}$$

which implies

$$\tilde{z}(\xi) = \tilde{z}(y_k) \text{ for } \xi \in [y_k, x_k] \text{ and } \tilde{z}(\eta) \leq z(\eta) \text{ for } \eta \in (x_k, y_{k+1}). \qquad (4.62)$$

For $k = 1, \ldots, m - 1$ we define x_k, y_{k+1} by

$$x_k := \min \{x \mid x > y_k \text{ and } \tilde{z}(y_k) = z(x)\},$$
$$y_{k+1} := \min \{y \mid y > x_k \text{ and } \tilde{z}(y) = z(x_k) - \gamma\}.$$

An obvious consequence from this definition is for $k = 1, \ldots, m - 1$

$$z(y_{k+1}) - \gamma \leq z(x_k) - \gamma = \tilde{z}(y_{k+1}) \leq z(y_{k+1}) \qquad (4.63)$$

by the assumed monotonicity of z. Note, that the last inequality holds by (4.62). Furthermore we define x_m in this way, such one can reach the point ξ_2 with slope $-M$, which usually yields $\tilde{z}(x_m) < z(x_m)$.

All in all we define

$$\tilde{z}(x) := \left\{ \begin{array}{ll} z_n, & \text{if } x \in [0,1] \backslash (\xi_1, \xi_2) \\ \tilde{z}, & \text{elsewhere.} \end{array} \right.$$

Furthermore we have

$$\tilde{z}_{|(\xi_1, \xi_2)} < z_{|(\xi_1, \xi_2)}, \ ||z - \tilde{z}||_{|(\xi_1, \xi_2)} < \gamma, \ (z_n - \tilde{z})_{|(\xi_1, \xi_2)} < -\gamma. \qquad (4.64)$$

Note, that $(4.64)_2$ holds by virtue of (4.63). We obtain by (4.60)

$$\hat{W}(x, \tilde{z}(x)) < \hat{W}(x, z_n(x)) \qquad (4.65)$$

for every $x \in (\xi_1, \xi_2)$.

It remains to show

$$\int_{\xi_1}^{\xi_2} \tilde{z}'^2 \, dx < \int_{\xi_1}^{\xi_2} z_n'^2 \, dx : \qquad (4.66)$$

We have

$$\int_{\xi_1}^{\xi_2} \tilde{z}'^2 \, dx = \sum_{k=1}^{m} \int_{x_k}^{y_{k+1}} M^2 \, dx.$$

For $k = 1, \ldots, m - 1$ we obtain $y_{k+1} - x_k = \frac{\gamma}{M}$ and for $k = m$ we get $y_{m+1} - x_m \leq \frac{\gamma}{M}$ which immediately yields

$$\int_{\xi_1}^{\xi_2} \tilde{z}'^2 \, dx \leq mM\gamma. \qquad (4.67)$$

Furthermore by a rough estimate we obtain by virtue of (4.59)

$$\int_{\xi_1}^{\xi_2} z_n'^2 \, dx \geq \frac{(z_n(\xi_1) - z_n(x_n))^2}{x_n - \xi_1} + \frac{(z_n(\xi_2) - z_n(x_n))^2}{\xi_2 - x_n} \geq$$
$$\geq 2 \frac{(\nu_M - \nu_r)^2}{\xi_2 - \xi_1} \geq 2 \frac{(\nu_M - \nu_r)^2}{\beta - \alpha}.$$

The following inequality is obvious:

$$\frac{z(\xi_2) - z(\xi_1)}{\xi_2 - \xi_1} \leq M$$

and hence we conclude

$$\frac{m\gamma}{\beta - \alpha} \leq M.$$

By (4.67) we obtain

$$\int_{\xi_1}^{\xi_2} \tilde{z}'^2 \, dx \leq M^2 (\beta - \alpha).$$

Thus (4.66) is valid, if the inequality

$$M^2(\beta - \alpha) \leq \left(\frac{||b||_{\infty,\Omega}}{\min\{\sigma'(\nu_m), \sigma'(\nu_M)\}} \right)^2 (\beta - \alpha) \leq 2 \frac{(\nu_M - \nu_r)^2}{\beta - \alpha}$$

is satisfied, which is true by virtue of (4.58) and (4.61). □

Remark 78. Note, that by (4.59) we have the following (without assuming (4.58)): Let z_ε be a sequence of global minimizers of (P_ε) and let $\Omega \subseteq [0, 1]$ be a nodal domain where no jump of z, the global minimizer of (P), occurs. Then z_ε does not cross the interval $[\nu_m, \nu_M]$ in Ω.

Furthermore we obtain by Lemma 77 an essential tool for proving uniform convergence of $(z_\varepsilon)_\varepsilon$:

Corollary 79. *Let $(z_\varepsilon)_\varepsilon$ be a sequence of stable stable solutions of* (4.17) *where b satisfies either property (H–2) or $(z_\varepsilon)_\varepsilon$ is a sequence of global minimizers, b satisfies (H–1) and let* (4.58) *be valid. Moreover let x_j, $j = 1, \ldots, r$ be the points of discontinuity of the global minimizer z of* (P) *and let $x_0 = 0$ and $x_{r+1} = 1$. Furthermore let $\gamma > 0$ be fixed but arbitrary. Then we obtain for every $\varepsilon < \varepsilon_0$, for arbitrary $\varrho > 0$ and for all $x \in [x_j + \gamma, x_{j+1} - \gamma]$, $j = 0, \ldots, r$*

$$z_\varepsilon(x) \begin{cases} > \nu_M - \varrho & , \text{if } z(x) > \nu_M \\ < \nu_m + \varrho & , \text{if } z(x) < \nu_m. \end{cases}$$

Proof. Let $\gamma > 0$ fixed but arbitrary and we consider the interval $[x_j + \gamma, x_{j+1} - \gamma] =: J_\gamma$ for some $j \in \{0, \ldots, r\}$. By virtue of Lemma 71 and 77 we immediately obtain the assertion for $J_\gamma \cap \hat{\Omega}$, where $\hat{\Omega}$ is defined by (4.44). Without loss of generality we assume

$$\int_x^1 b(\tau)d\tau + t_r > \sigma(\nu_M) \quad \text{for every } x \in J_\gamma, \tag{4.68}$$

which implies $z(x) > \nu_M$, where z denotes the global minimizer of (P).
Let $a := x_j + \gamma$ and $[a, x_{j+1} - \gamma] \cap A_j = [a, x_j^2]$, A_j defined by (4.10). By the pointwise convergence of $(z_\varepsilon)_\varepsilon$ to the global minimizer of (P) we obtain

$$z_\varepsilon(a) \geq \nu_M - \frac{\varrho}{2}$$

for an arbitrary $\varrho > 0$ and $\varepsilon < \varepsilon_0$. Thus it remains to show $z_\varepsilon(x) \geq \nu_M - \varrho$ for every $x \in [a, x_j^2]$:
By virtue of Lemma 69 we know $z_\varepsilon(x_j^2) > \nu_M$ for $\varepsilon < \varepsilon_0$ and furthermore Theorem 80 yields that z_ε crosses $[\nu_m, \nu_M]$ strictly monotonically increasing in A_j. If we assume for

a moment the existence of z_ε and $x \in [a, x_j^2]$ such that $z_\varepsilon(x) < \nu_M - \varrho$ we obtain a nodal domain $[x', x'']$ of z_ε' with $z_{\varepsilon|[x',x'']}' < 0$. Due to (4.68) we have $b_{|A_j} < 0$ and thus

$$\text{sign}(bz_\varepsilon')_{|[x',x'']} > 0$$

contradicting the stability of z_ε.   □

The qualitative behavior of a sequence $(z_\varepsilon)_\varepsilon$ satisfying property (E) concerning the crossing of $[\nu_m, \nu_M]$ is given by the following theorem:

Theorem 80. *Let $(z_\varepsilon)_\varepsilon$ satisfies property (E). Then there exists ε_0 sufficiently small such that for every $\varepsilon < \varepsilon_0$ the following is valid: The function z_ε crosses $[\nu_m, \nu_M]$ exactly r–times in the interior of the intervals A_j, $j = 1, \ldots, r$. Moreover z_ε is strictly monotonically during this crossing.*

Proof. By Lemma 69 z_ε crosses $[\nu_m, \nu_M]$ in the interior of A_j, $j = 1, \ldots, r$, which yields that z_ε crosses the interval at least r–times and we will show the claimed strict monotonicity:
We consider a certain $A_j = [x_1^j, x_2^j]$ and without loss of generality we assume $b_{|A_j} > 0$ which implies

$$x \mapsto \left(\int_x^1 b(\tau) d\tau + t_r \right)_{|A_j} \quad \text{is strictly monotonically decreasing.}$$

Again by Lemma 69 we obtain

$$z_\varepsilon(x_1^j) > \nu_M \qquad \text{and} \qquad z_\varepsilon(x_2^j) < \nu_m. \tag{4.69}$$

We define

$$\alpha := \max \left\{ x \in A_j \mid z_\varepsilon(x) = \nu_m \right\},$$
$$\beta := \max \left\{ x \in A_j \mid z_\varepsilon(x) = \nu_M \right\},$$

and $\alpha, \beta \in \overset{\circ}{A}_j$ is obvious, where $\overset{\circ}{A}_j$ denotes the interior of A_j. Furthermore we have $z_\varepsilon'(\gamma) < 0$ for $\gamma = \{\alpha, \beta\}$ which can be established as follows:
We assume for a moment $z_\varepsilon'(\gamma) = 0$ for $\gamma \in \{\alpha, \beta\}$ (note, that ">" can be easily excluded by the definition of α and β). Because any zero of z_ε' is non–degenerate by Theorem 66, z_ε' changes sign at $x = \gamma$ from $+$ to $-$. This fact together with (4.69) implies the existence of a nodal domain $V \subseteq A_j$ of z_ε' such that $z_{\varepsilon|V}' > 0$ and in particular we have

$$(z_\varepsilon' b)_{|V} > 0,$$

which contradicts stability of z_ε. By the same procedure one can show that $z_{\varepsilon|[\alpha,\beta]}' < 0$, which proves strict monotonicity during the crossing.
It remains to show that z_ε crosses $[\nu_m, \nu_M]$ exactly r–times. By Lemma 71 respectively Remark 78 we know, z_ε cannot cross $[\nu_m, \nu_M]$ in $\hat{\Omega} = [0, 1] \setminus \bigcup_{j=1}^r A_j$. Therefore we assume the existence of a certain A_j in which z_ε crosses $[\nu_m, \nu_M]$ at least two times and without loss of generality we consider again the case $b_{|A_j} > 0$. Because of (4.69) and our assumption we obtain at least three crossings in A_j contradicting again stability of z_ε.   □

We conclude this section with a corollary, which is an easy consequence of Theorem 80:

Corollary 81. *Let $(z_\varepsilon)_\varepsilon$ be a sequence satisfying property (E). Then for $\varepsilon < \varepsilon_0$ the following is valid: For each A_j, $j = 1, \ldots, r$ and for each $\nu \in [\nu_m, \nu_M]$ there exists exactly one $\alpha_j(\varepsilon) \in A_j$ with*

$$z_\varepsilon(\alpha_j) = \nu.$$

4.5 Asymptotic Behavior

In this section we consider the asymptotic behavior of a sequence $(z_n)_{n \in \mathbb{N}}$ of global minimizers respectively a sequence satisfying property (E). In particular, we study the width of an interval depending on ε in which the crossing from ν_m to ν_M or vice versa takes place. Furthermore we prove uniform convergence of $(z_n)_{n \in \mathbb{N}}$ at least in certain areas with a convergence–rate also depending on ε.

Let $x_j \in A_j$ be a jump point of the global minimizer z of (P). Furthermore we know by Theorem 80 that every global minimizer z_ε of (P_ε) crosses the interval $[\nu_m, \nu_M]$ strictly monotonically in the interior of A_j. For fixed but arbitrary $\delta > 0$ we deduce certain points $x_{\nu_m}^\varepsilon$ and $x_{\nu_M}^\varepsilon \in A_j$, uniquely defined by the mentioned monotonicity, such that

$$z_\varepsilon(x_{\nu_m}^\varepsilon) = \nu_m + \delta \qquad \text{and} \qquad z_\varepsilon(x_{\nu_M}^\varepsilon) = \nu_M - \delta. \tag{4.70}$$

For the width of the transition layer we have the following result:

Lemma 82. *Let z_ε be the global minimizer of (P_ε) and let b satisfy (H-1). Furthermore let $x_{\nu_m}^\varepsilon$ and $x_{\nu_M}^\varepsilon$ be defined as above. Then*

$$\left|\left| x_{\nu_M}^\varepsilon - x_{\nu_m}^\varepsilon \right|\right| \leq C\sqrt{\varepsilon} \tag{4.71}$$

holds with a constant $C > 0$ just depending on δ.

Proof. Without loss of generality we assume $x_{\nu_m}^\varepsilon < x_{\nu_M}^\varepsilon$. Let z, z_ε be the global minimizer of (P) respectively (P_ε) . Equation (4.54) yields

$$\hat{W}(x, z_\varepsilon(x)) - \hat{W}(x, z(x)) \geq 0$$

for every $x \in [0, 1]$ and by (4.30) we obtain

$$\int_{x_{\nu_m}^\varepsilon}^{x_{\nu_M}^\varepsilon} \left(\hat{W}(x, z_\varepsilon(x)) - \hat{W}(x, z(x)) \right) dx =$$

$$\int_0^1 \left(\hat{W}(x, z_\varepsilon(x)) - \hat{W}(x, z(x)) \right) dx - \left(\int_0^{x_{\nu_m}^\varepsilon} + \int_{x_{\nu_M}^\varepsilon}^1 \right) \left(\hat{W}(x, z_\varepsilon(x)) - \hat{W}(x, z(x)) \right) dx \leq$$

$$\int_0^1 \left(\hat{W}(x, z_\varepsilon(x)) - \hat{W}(x, z(x)) \right) dx \leq K\sqrt{\varepsilon}.$$

$$\tag{4.72}$$

Moreover we claim for an arbitrary $\delta > 0$ the existence of a constant $C = C(\delta) > 0$ such that for all $\varepsilon > 0$ and for every $x \in [x_{\nu_m}^\varepsilon, x_{\nu_M}^\varepsilon]$ we have

$$\hat{W}(x, z_\varepsilon(x)) - \hat{W}(x, z(x)) \geq C(\delta) : \tag{4.73}$$

By virtue of Theorem 80 we get $z_\varepsilon(x) \in [\nu_m + \delta, \nu_M - \delta] := J_\delta$ for every $x \in [x_{\nu_m}^\varepsilon, x_{\nu_M}^\varepsilon]$ and this yields

$$\hat{W}(x, z_\varepsilon(x)) \geq \min_{\alpha \in J_\delta} \hat{W}(x, \alpha).$$

Trivially it holds for every $x \in [x_{\nu_m}^\varepsilon, x_{\nu_M}^\varepsilon]$

$$\hat{W}(x, z_\varepsilon(x)) - \hat{W}(x, z(x)) \geq \min_{\alpha \in J_\delta} \hat{W}(x, \alpha) - \hat{W}(x, z(x)). \tag{4.74}$$

To confirm (4.73) it is enough to prove

$$\forall \delta > 0 \, \exists C > 0 \, \forall x \in [0,1] : \min_{\alpha \in J_\delta} \hat{W}(x, \alpha) - \hat{W}(x, z(x)) \geq C. \tag{4.75}$$

We assume for a moment the contrary of (4.75) and hence there exists a certain $\delta > 0$ such that for $\frac{1}{n}$, $n \in \mathbb{N}$, there exists $x_n \in [0,1]$ with

$$\min_{\alpha \in J_\delta} \hat{W}(x_n, \alpha) - \hat{W}(x_n, z(x_n)) \leq \frac{1}{n}.$$

A subsequence of $(x_n)_{n \in \mathbb{N}}$ (not relabelled) converges to $x_0 \in [0,1]$ and this yields

$$\min_{\alpha \in J_\delta} \hat{W}(x_0, \alpha) = \hat{W}(x_0, z(x_0)). \tag{4.76}$$

Note, that $x \mapsto \hat{W}(x, z(x))$ is continuous by Lemma 58 (iv). But this yields a contradiction because the global minimum of $s \mapsto \hat{W}(x_0, s)$ is never achieved in $[\nu_m + \delta, \nu_M - \delta]$ by Propositions 72 and 73. By (4.72) and (4.73) we deduce

$$C(\delta) \left|\left| x_{\nu_M}^\varepsilon - x_{\nu_m}^\varepsilon \right|\right| \leq \int_{x_{\nu_m}^\varepsilon}^{x_{\nu_M}^\varepsilon} \left(\hat{W}(x, z_\varepsilon(x)) - \hat{W}(x, z(x)) \right) dx \leq K\sqrt{\varepsilon}.$$

Division by $C(\delta)$ yields the result. $\square$

Remark 83. One can find a similar proof in [4]. But unfortunately their proof seems to be incomplete.

In the following we address the question of uniform convergence of a sequence $(z_n)_{n \in \mathbb{N}}$ of stable solutions of (4.17). Our first result in this direction concerns the convergence–rate with respect to ε under the assumption of uniform convergence. The proof borrows on ideas of [2], Lemma 2.2.

Lemma 84. *Let $(z_\varepsilon)_\varepsilon$ be a sequence of stable solutions of* (4.17) *and let z be the global minimizer of* (P). *We assume $z_\varepsilon \to z$ uniformly in an interval $[a,b]$ for $\varepsilon \searrow 0$. If $[c,d] \subseteq (a,b)$ is fixed, then we obtain*

$$||z_\varepsilon - z||_{C^0(c,d)} \leq C\sqrt{\varepsilon}, \tag{4.77}$$

where $C > 0$ denotes a constant independent of ε.

Proof. Let $(z_\varepsilon)_\varepsilon$ be a sequence of stable solutions of (4.17) converging uniformly to z (the minimizer of (P)) in $[a,b]$ and choose $[c,d] \subseteq (a,b)$. Furthermore we fix $\bar{c} \in (a,c)$, $\bar{d} \in (d,b)$ and by the mean–value theorem we obtain $a_\varepsilon \in (a,\bar{c})$ and $b_\varepsilon \in (\bar{d},b)$, such that

$$z'_\varepsilon(a_\varepsilon) = \frac{z_\varepsilon(\bar{c}) - z_\varepsilon(a)}{\bar{c} - a},$$

$$z'_\varepsilon(b_\varepsilon) = \frac{z_\varepsilon(b) - z_\varepsilon(\bar{d})}{b - \bar{d}}$$

holds. This implies immediately by (4.22)

$$||z'_\varepsilon(a_\varepsilon)|| \leq C \quad \text{and} \quad ||z'_\varepsilon(b_\varepsilon)|| \leq C \tag{4.78}$$

with a constant C independent of ε.
Furthermore we have due to (4.13) and (4.17)

$$-\varepsilon z'' + \sigma(z_\varepsilon) = \int_x^1 b(\tau)d\tau + t_r = \sigma(z) \tag{4.79}$$

for every $x \in (0,1)$. Multiplying (4.79) by $z_\varepsilon - z$ and integration by parts yields

$$\begin{aligned}
0 &= \int_{a_\varepsilon}^{b_\varepsilon} -\varepsilon z''_\varepsilon(z_\varepsilon - z) + (\sigma(z_\varepsilon) - \sigma(z))(z_\varepsilon - z)\, dx = \\
&= \int_{a_\varepsilon}^{b_\varepsilon} \varepsilon z'_\varepsilon(z_\varepsilon - z)' + (\sigma(z_\varepsilon) - \sigma(z))(z_\varepsilon - z)\, dx - \varepsilon z'_\varepsilon(z_\varepsilon - z)\,|_{a_\varepsilon}^{b_\varepsilon}.
\end{aligned} \tag{4.80}$$

The assumed uniform convergence implies either $z_{|[a,b]} \in [\nu_M, \infty]$ or $z_{|[a,b]} \in [-\infty, \nu_m]$. Moreover we have for fixed but arbitrary small $\varrho > 0$ and for $\varepsilon < \varepsilon_0$ the estimate $||z_\varepsilon - z||_{C^0[a,b]} < \varrho$. By these two facts we deduce

$$\sigma'(z_\varepsilon(x)) \geq C_1 > 0 \tag{4.81}$$

for every $x \in [a,b]$ and for all $\varepsilon < \varepsilon_0$. Therefore we get due to (4.81) and the mean–value theorem

$$\begin{aligned}
&\int_{a_\varepsilon}^{b_\varepsilon} (\sigma(z_\varepsilon) - \sigma(z))(z_\varepsilon - z)\, dx = \\
&\int_{a_\varepsilon}^{b_\varepsilon} \sigma'(\xi(x))(z_\varepsilon - z)^2\, dx \geq C_1 \int_{a_\varepsilon}^{b_\varepsilon} (z_\varepsilon - z)^2\, dx
\end{aligned}$$

with a certain $\xi(x) \in [z_\varepsilon(x), z(x)]$ and (4.80) yields

$$0 \geq \int_{a_\varepsilon}^{b_\varepsilon} \varepsilon z'^2_\varepsilon - \varepsilon z'_\varepsilon z' + C_1(z_\varepsilon - z)^2\, dx - \varepsilon z'_\varepsilon(z_\varepsilon - z)\,|_{a_\varepsilon}^{b_\varepsilon}.$$

This implies

$$\begin{aligned}
\int_{a_\varepsilon}^{b_\varepsilon} C_1(z_\varepsilon - z)^2\, dx &\leq \int_{a_\varepsilon}^{b_\varepsilon} \varepsilon ||z'_\varepsilon z'||\, dx + \varepsilon z'_\varepsilon(z_\varepsilon - z)\,|_{a_\varepsilon}^{b_\varepsilon} \leq \\
&\leq \varepsilon ||z'||_{C^0(a_\varepsilon, b_\varepsilon)} ||z'_\varepsilon||_{L^1(a_\varepsilon, b_\varepsilon)} + M\varepsilon.
\end{aligned} \tag{4.82}$$

Note, that $z'_\varepsilon(z_\varepsilon - z)\,|_{a_\varepsilon}^{b_\varepsilon}$ can be bounded independently of ε by virtue of (4.78). Furthermore we have by the fundamental theorem in calculus the estimate $||z'_\varepsilon||_{L^1(0,1)} \leq K$ also with K independent of ε and due to

$$z'(x) = \frac{d}{dx}\left(\sigma^{-1}\right)\left(\int_x^1 b(\tau)d\tau + t_r\right) \tag{4.83}$$

for every $x \in [a,b]$ we get $||z'||_{C^0(a_\varepsilon,b_\varepsilon)} \leq C$. Thus (4.82) yields

$$\int_{a_\varepsilon}^{b_\varepsilon} (z_\varepsilon - z)^2\, dx \leq M\varepsilon. \tag{4.84}$$

We have $[c,d] \subseteq [\bar{c},\bar{d}] \subseteq [a_\varepsilon, b_\varepsilon]$ and by (4.84) it follows the existence of points $c_\varepsilon \in (\bar{c},c)$ and $d_\varepsilon \in (d,\bar{d})$ such that

$$||z_\varepsilon(c_\varepsilon) - z(c_\varepsilon)|| \leq C\sqrt{\varepsilon} \qquad \text{and} \qquad ||z_\varepsilon(d_\varepsilon) - z(d_\varepsilon)|| \leq C\sqrt{\varepsilon} \tag{4.85}$$

with a certain $C > 0$ independent of $\varepsilon < \varepsilon_0$.

Furthermore we know after differentiating (4.83) (which is possible because b and σ are assumed to be sufficiently smooth), that $z''_{|[a,b]}$ exists and

$$||z''||_{C^0(a,b)} \leq \hat{K}. \tag{4.86}$$

To finish the proof we have to show the following:

$$\exists K > 0\, \exists \varepsilon_0 > 0 \,\forall \varepsilon < \varepsilon_0 \,\forall x \in [c_\varepsilon, d_\varepsilon] : ||z(x) - z_\varepsilon(x)|| \leq K\sqrt{\varepsilon}.$$

Suppose not: Then we obtain for $K := \max\left\{\frac{\hat{K}}{C_1}, C + 1\right\}$ a certain $\varepsilon < \varepsilon_0$ and $x_0 \in [c_\varepsilon, d_\varepsilon]$ such that

$$||z(x_0) - z_\varepsilon(x_0)|| > K\sqrt{\varepsilon}. \tag{4.87}$$

We have to distinguish two cases:

(a) We assume that $z - z_\varepsilon$ possesses a global positive maximum at $x_0 \in (c_\varepsilon, d_\varepsilon)$:
    By virtue of $K > C$ and (4.85) we obtain $x_0 \notin \{c_\varepsilon, d_\varepsilon\}$ and thus $z - z_\varepsilon$ has a positive maximum in the interior of $[c_\varepsilon, d_\varepsilon]$. By (4.13) and (4.17) we deduce the following equality:

$$\varepsilon(z - z_\varepsilon)'' - (\sigma(z) - \sigma(z_\varepsilon)) = \varepsilon z''. \tag{4.88}$$

In particular this yields for $x = x_0$ by (4.81) and (4.86):

$$\varepsilon(z - z_\varepsilon)''(x_0) = \varepsilon z''(x_0) + (\sigma(z(x_0)) - \sigma(z_\varepsilon(x_0))) \geq$$
$$\geq -\varepsilon\hat{K} + C_1 K\sqrt{\varepsilon} = \sqrt{\varepsilon}(C_1 K - \sqrt{\varepsilon}\hat{K}) \geq \sqrt{\varepsilon}(\hat{K} - \sqrt{\varepsilon}\hat{K}) > 0,$$

contradicting our assumption.

(b) Let $z - z_\varepsilon$ possess a negative minimum at $x_0 \in (c_\varepsilon, d_\varepsilon)$: A similar proof as in (a) yields a contradiction and the proof is done.

$$\square$$

Our next goal is to determine the regions in $(0, 1)$ in which a sequence of stable solutions of the Euler–Lagrange equation (4.17) will converge uniformly to the global minimizer of (P).

Theorem 85. *Let $(z_\varepsilon)_\varepsilon$ be a sequence with the same properties as the one in Corollary 79. Furthermore let x_j, $j = 1, \ldots, r$ be the jump points of the global minimizer z of (P) and let $x_0 = 0$ and $x_{r+1} = 1$. Then we obtain*

$$\lim_{\varepsilon \to 0} z_\varepsilon = z$$

uniformly on $T_j := (x_j + 2\delta, x_{j+1} - 2\delta)$ for every $j = 0, \ldots, r$ and for fixed but arbitrary $\delta > 0$. In particular we obtain by virtue of Lemma 84

$$||z_\varepsilon - z||_{\infty, T_j} \leq K \sqrt{\varepsilon}$$

for every $j = 0, \ldots, r$.

Proof. Let $(z_n)_{n \in \mathbb{N}}$ be some sequence with the same properties as the one considered in Corollary 79 and let z be the global minimizer of (P). We assume the existence of a certain T_j such that $(z_n)_{n \in \mathbb{N}}$ does not converge uniformly to z in T_j. Without loss of generality we assume $z_{|T_j} > \nu_M$ and our assumption implies

$$\exists \varrho > 0 \, \forall n_0 \, \exists n > n_0 \, \exists x_n \in T_j : ||z_n(x_n) - z(x_n)|| > \varrho.$$

Without loss of generality we assume

$$z_n(x_n) - z(x_n) > \rho \tag{4.89}$$

for every $n \in \mathbb{N}$ and furthermore we have $\lim_{n \to \infty} x_n = x_0 \in \bar{T}_j$.

Let $a < x_0$ and $b > x_0$ and due to the pointwise convergence of $(z_n)_{n \in \mathbb{N}}$ to the global minimizer of (P) we obtain

$$||z_n(a) - z(a)|| < \frac{\varrho}{2} \quad \text{and} \quad ||z_n(b) - z(b)|| < \frac{\varrho}{2} \tag{4.90}$$

for every $n \in \mathbb{N}$.

We define $w_n := z_n - z$ and by the mean–value theorem and (4.89) there exists $c_n, d_n \in (a, b)$ such that

$$w_n(c_n) = w_n(d_n) = \frac{\varrho}{2}, \quad w_{n|(c_n, d_n)} > \frac{\varrho}{2}$$

and furthermore we have $x_n \in (c_n, d_n)$.

This implies immediately the existence of a global positive maximum of w_n in (c_n, d_n). By virtue of Corollary 79 we assume

$$z_{n|T_j} > \nu_M - \varrho \text{ for a certain } \varrho > 0 \text{ and for every } n \in \mathbb{N}.$$

Due to the Euler–Lagrange equation we have for every $n \in \mathbb{N}$ and for all $x \in (c_n, d_n)$

$$\varepsilon_n w_n'' = \varepsilon_n(z_n'' - z'') = \sigma(z_n) - \int_x^1 b(\tau)d\tau - t_r - \varepsilon_n z'' =$$
$$= \sigma(z_n) - \sigma(z) - \varepsilon_n z'' = \sigma'(\xi)(z_n - z) - \varepsilon_n z''.$$

Because of $\sigma'(\xi(x)) \geq K > 0$ and $z''(x) \leq \tilde{K}$ for every $x \in (x_j + \delta, x_{j+1} - \delta)$ the previous equation yields

$$\varepsilon_n w_n'' \geq K\frac{\varrho}{2} - \tilde{K}\varepsilon_n > 0$$

for n sufficiently large. Therefore we have on the one hand that w_n has a global positive maximum in (c_n, d_n) and on the other hand $w''_{n|(c_n,d_n)} > 0$, hence a contradiction.   □

4.6   Necessary condition for uniqueness of the global Minimizer

In this section we give a necessary condition to ensure uniqueness of the global minimizer of (P_ε) for ε sufficiently small in terms of the position of the transition layer provided the body–force b is either positive or negative. From a physical point of view this is essentially the case, if we consider a gravitational body–force ($b > 0$). To deduce this result, we follow the lines of [6]. A similar statement was given in [4] also proved by the method proposed in [6].
We consider the variational problem (P_ε) in the following form

$$J(z) := \int_0^1 \left(\frac{\varepsilon}{2}z'^2 + W(z) - \int_x^1 b(\tau)d\tau z - t_r z \right) dx$$

and we assume the existence of two global minimizers z_1, $z_2 \in W^{1,2}(0,1)$ of (P_ε). A crucial observation for our forthcoming analysis is the following lemma which was proved in [4]. The corresponding result for the Dirichlet problem was established in [16].

Lemma 86. *Global minimizers of (P_ε) are strictly pointwise ordered, i.e. given z_1, $z_2 \in W^{1,2}(0,1)$, both global minimizers of (P_ε), we have either $z_1(x) < z_2(x)$ or $z_2(x) < z_1(x)$ for every $x \in [0,1]$.*

For the proof we refer to [4], Lemma 5 in Chapter 5.
First let us give the precise assumptions on b:

(i) The body–force b is either positive or negative. (For the sequel of the section we assume $b > 0$).

(ii) We have the existence of a unique $x_1 \in (0,1)$ such that $\int_{x_1}^1 b(\tau)d\tau + t_r = \sigma(\nu_m)$.

Let z be the global minimizer of (P) and due to our conditions on b we get exactly one jump of z at $x = x_1$. Furthermore let z_1^n, z_2^n be two global minimizers of (P_{ε_n}) and by Lemma 86 we obtain (without loss of generality) $z_1^n < z_2^n$. We assume the existence of an arbitrary $z_n \in [z_1^n, z_2^n]$ (not necessarily stable), such that

$$-\varepsilon_n z_n'' + \sigma(z_n) = \int_x^1 b(\tau)d\tau + t_r,$$
$$z_n'(0) = z_n'(1) = 0$$

(4.91)

is fulfilled. By assumption (ii) on b, Theorem 80 and $z_n \in [z_1^n, z_2^n]$ we can define for fixed but arbitrary $\delta > 0$ and every $n \in \mathbb{N}$

$$a_n := \min \{x \mid z_n(x) = \nu_M - \delta\},$$
$$b_n := \max \{x \mid z_n(x) = \nu_m + \delta\}, \tag{4.92}$$

such that $a_n < b_n$. Moreover we have $\sigma'(z_n)_{|(0,a_n)\cup(b_n,1)} \geq k > 0$.

With respect to z_1^n, z_2^n we get (unique) points $a_j^n, b_j^n, j = 1, 2$, defined as a_n, b_n in (4.92) and due to $z_1^n < z_2^n$ we obtain $a_1^n < a_2^n < b_2^n$.

In the sequel we assume the following inequality which is essential for our forthcoming analysis:

$$||a_1^n - b_2^n|| \leq C\sqrt{\varepsilon_n} \quad \text{for every } n \in \mathbb{N}. \tag{4.93}$$

Remark 87. Unfortunately we were not able to prove (4.93). But if we assume the existence of two global minimizers of (P_ε) and (4.93) holds, then our proof will provide us a contradiction. Hence, the global minimizer of (P_ε) is unique.

Furthermore there exists (a not necessarily unique point) $x_c^n \in (a_n, b_n)$ such that

$$z_n(x_c^n) = \frac{\nu_m + \nu_M}{2}. \tag{4.94}$$

The function Z_n is defined by

$$Z_n(t) := z_n(a_n + \sqrt{\varepsilon_n}t) \tag{4.95}$$

for every $t \in \left(-\frac{a_n}{\sqrt{\varepsilon}}, \frac{1-a_n}{\sqrt{\varepsilon}}\right) =: J_n$. Due to (4.91) Z_n satisfies the following equation

$$-Z_n''(t) + \sigma(Z_n(t)) = \int_{a_n+\sqrt{\varepsilon_n}t}^1 b(\tau) \, d\tau + t_r. \tag{4.96}$$

Moreover we have by virtue of $||z_n||_\infty \leq C$ for every $n \in \mathbb{N}$ the uniform estimate $||Z_n||_{C_0(J_n)} \leq C$. This implies by (4.96) $||Z_n''||_{C_0(J_n)} \leq C$ for all $n \in \mathbb{N}$ and interpolation yields

$$||Z_n||_{C^2(J_n)} \leq C \quad \text{for every } n \in \mathbb{N}. \tag{4.97}$$

Therefore we have for $[-m, m]$ with arbitrary $m \in \mathbb{N}$ the estimate $||Z_n||_{C^2([-m,m])} \leq C$ for every $n > n_0$ such that $[-m, m] \subseteq J_n$. Due to the compact embedding $C^2([-m,m]) \hookrightarrow C^1([-m,m])$ we obtain

$$Z_n \to Z \quad \text{in } C^1([-m,m]).$$

Furthermore we claim $(Z_n'')_{n\in\mathbb{N}}$ is relative compact in $C^0([-m,m])$ for arbitrary $m \in \mathbb{N}$:

$$\forall \varrho > 0 \exists \delta > 0 \forall t_1, t_2 \in [-m, m] \forall (Z_n)_{n>n_0} : ||t_1 - t_2|| < \delta \Rightarrow ||Z_n''(t_1) - Z_n''(t_2)|| < \varrho.$$

Due to $||Z_n||_{C^0[-m,m]} \leq C$ and (4.96) there exists K_1, K_2 such that

$$||\sigma'(Z_n)||_{C^0([-m,m])} \leq K_1 \quad \text{and} \quad ||Z_n'||_{C^0([-m,m])} \leq K_2 \tag{4.98}$$

for every $n \in \mathbb{N}$ and furthermore we have $||b||_\infty \leq K_3$. We define $\delta := \frac{\varrho}{K_1 K_2 + K_3}$ and therefore we have for $||t_1 - t_2|| < \delta$ and arbitrary n by (4.96)

$$||Z_n''(t_1) - Z_n''(t_2)|| \leq ||\sigma(Z_n(t_1)) - \sigma(Z_n(t_2))|| + \left\| \int_{a_n + \sqrt{\varepsilon_n} t_2}^{1} b \, d\tau - \int_{a_n + \sqrt{\varepsilon_n} t_1}^{1} b \, d\tau \right\| \leq$$

$$\leq K_1 ||Z_n(t_1) - Z_n(t_2)|| + ||b||_\infty \sqrt{\varepsilon_n} ||t_2 - t_1|| \leq (K_1 K_2 + K_3) ||t_2 - t_1|| < \varrho.$$

This yields the following:

Lemma 88. *Let $m \in \mathbb{N}$ be arbitrary but fixed. Then the sequence $(Z_n)_{n \in \mathbb{N}}$ is relative compact in $C^2([-m, m])$. Therefore there exists Z such that $Z_n \to Z$ in $C^2([-m, m])$. We denote this by $Z_n \to Z$ in $C^2_{loc}(\mathbb{R})$.*
In particular, Z satisfies

$$Z''(t) - \sigma(Z(t)) = - \int_{x_1}^{1} b(\tau) d\tau - t_r = -\sigma(\nu_m) \tag{4.99}$$

for every $t \in \mathbb{R}$.

The next Lemma states an important property of the limit Z.

Lemma 89. *Let Z be the limit obtained by virtue of Lemma 88. Then Z is strictly monotonically decreasing in $\mathbb{R}$.*

Proof. Let z_1^n, z_2^n be two global minimizers of (P_{ε_n}) and let $z_n \in [z_1^n, z_2^n]$ be a solution of (4.91). We seperate the proof into four steps:

Step 1:   First we prove $z_{n|[0,a_n]}$ is strict monotonically decreasing:
Because of $b > 0$ we deduce due to stability of global minimizers

$$(z_2^n)'(x) < 0 \quad \text{for every } x \in [0, 1]. \tag{4.100}$$

Furthermore for all $x \in [0, a_n]$ we have due to $\nu_M - \delta \leq z_n(x) \leq z_2^n(x)$ the estimate $\sigma(z_n(x)) \leq \sigma(z_2^n(x))$. Moreover the following equation is valid

$$-\varepsilon_n z_n''(x) + \sigma(z_n(x)) = -\varepsilon_n (z_2^n)''(x) + \sigma(z_2^n(x)).$$

This implies

$$\varepsilon_n (z_2^n - z_n)''(x) \geq 0 \tag{4.101}$$

for every $x \in [0, a_n]$. If we assume $(z_2^n - z_n)''(0) = 0$ we deduce immediately by standard ODE theory $(z_2^n)' \equiv z_n'$ on the unit interval and the proof is done by virtue of (4.100). Thus we can assume

$$\varepsilon_n (z_2^n - z_n)''(0) > 0. \tag{4.102}$$

We define $v_n := z_n - z_2^n \leq 0$. Because of the boundary condition $v_n'(0) = 0$ and $v_n''(0) < 0$ due to (4.102), we have

$$v_{n|(0,\gamma)}' < 0 \text{ for a certain } \gamma > 0. \tag{4.103}$$

Let us assume for a moment the existence of $x_1 \in [0, a_n]$ with

$$v_n'(x_1) = 0 \quad \text{and} \quad v_{n|(0,x_1)}' < 0.$$

The Euler–Lagrange equation (4.17) and the mean–value theorem implies with a certain $\xi(x) \in [z_n(x), z_2^n(x)] \subseteq [\nu_M - \delta, \infty)$

$$\begin{aligned} \varepsilon_n v_n'' - \sigma'(\xi)v_n &= 0, \\ v_n'(0) = v_n'(x_1) &= 0. \end{aligned} \tag{4.104}$$

Due to $\sigma'(\xi(x)) > 0$ we have $v_{n|[0,x_1)}$ is concave and hence

$$v_n'(x) \le v_n'(0) = 0 \quad \text{for every } x \in [0, x_1]$$

and moreover we have $v_n'(x_1) = 0 \Leftrightarrow v_n \equiv 0$.
Therefore we obtain either $z_n \equiv z_2^2$ or $v_{n|(0,a_n)}' < 0$. Both imply immediately $z_n'(x) < 0$ for every $x \in (0, a_n)$.

By step 1 we deduce $Z_{n|(-\frac{a_n}{\sqrt{\varepsilon_n}},0)}$ is strictly monotonically decreasing, which yields

$$Z_{|[-m,0]}' \le 0 \tag{4.105}$$

for an arbitrary $m \in \mathbb{N}$.

Step 2: In a similar manner one can prove $z_{n|(b_n,1)}$ is strictly monotonically decreasing and this implies

$$Z_{n|(\frac{b_n-a_n}{\sqrt{\varepsilon_n}}, \frac{1-a_n}{\sqrt{\varepsilon_n}})}' < 0.$$

With respect to (4.93) we have for every $n \in \mathbb{N}$ the inequality $0 < \frac{b_n-a_n}{\sqrt{\varepsilon_n}} < C$ and hence

$$Z_{|[C,m]}' \le 0 \tag{4.106}$$

for an arbitrary $m \in \mathbb{N}$.

Step 3: Z is monotonically decreasing in $\mathbb{R}$:
Suppose, there exists $[x_1, x_2] \subseteq [0, C]$ such that $Z_{|(x_1,x_2)}' > 0$ and $Z'(x_1) = Z'(x_2) = 0$.
We define for $x \in [0, x_2 - x_1]$

$$\tilde{Z}(x) := Z(x_2 - x)$$

and extend $\tilde{Z}$ by periodicity from $[0, x_2 - x_1]$ to the real line. Then $\tilde{Z}$ is the unique solution of the equation

$$\begin{aligned} \tilde{Z}''(x) - \sigma(\tilde{Z}(x)) &= Z''(x_2 - x) - \sigma(Z(x_2 - x)) = -\sigma(\nu_m), \\ \tilde{Z}(0) = Z(x_2) \quad &, \quad \tilde{Z}'(0) = -Z'(x_2) = 0, \end{aligned}$$

contradicting (4.105) and (4.106).

Step 4: Z is strictly monotonically decreasing in $\mathbb{R}$:
We assume, there exists $x_0 \in \mathbb{R}$ such that $Z'(x_0) = 0$ and $Z''(x_0) = 0$. Thus we have $\sigma(Z(x_0)) = \sigma(\nu_m)$ and this yields

$$Z(x_0) \in (\sigma)^{-1}(\nu_m) := \{\nu_m, \nu_r, \nu_M\}.$$

Furthermore Z satisfies the initial value problem

$$Z'' - \sigma(Z) = -\sigma(\nu_m),$$
$$Z(x_0) \in \{\nu_m, \nu_r, \nu_M\}, \quad Z'(x_0) = 0$$

and this implies $Z \equiv \{\nu_m, \nu_r, \nu_M\}$. On the other hand we have $Z(0) \equiv Z_n(0) = z_n(a_n) = \nu_M - \delta$. Contradiction. $\qquad\square$

Another property of Z is the following:

Lemma 90. *The function Z is the unique monotonically decreasing solution of*

$$Z'' - \sigma(Z) = -\sigma(\nu_m),$$
$$\lim_{t \to -\infty} Z(t) = \nu_M, \quad \lim_{t \to \infty} Z(t) = \nu_m, \quad Z(0) = \nu_M - \delta. \tag{4.107}$$

Proof. First note, that $Z(0) = \nu_M - \delta$ is obvious due to the fact, that

$$Z_n(0) = \nu_M - \delta \quad \text{for every } n \in \mathbb{N}.$$

Furthermore we know for arbitrary $m \in \mathbb{N}$

$$||Z_n||_{C^0(-m,m)} \leq ||z_n||_{C^0(0,1)} \leq K$$

with K independent of m and n. This yields immediately

$$||Z||_{C^0(\mathbb{R})} \leq K. \tag{4.108}$$

Moreover we have $Z' < 0$ and this implies

$$\lim_{t \to \pm\infty} Z(t) \text{ exists}$$

and by the Euler–Lagrange equation we obtain certain α, β with

$$\lim_{t \to -\infty} Z''(t) = \alpha, \quad \lim_{t \to \infty} Z''(t) = \beta.$$

We have to show $\alpha, \beta = 0$.
Obviously it is enough to prove this assertion for α, the other case is obtained in a similar manner:

(i) We assume $\alpha < 0$:
   This implies the existence of t_0 such that for every $t < t_0$ we obtain

$$Z''_{|(-\infty,t_0)} < \frac{\alpha}{2} < 0.$$

The inequality $\left|\left| Z'_{|[-m,m]} \right|\right| < C = C(m)$ for every $m \in \mathbb{N}$ yields

$$\lim_{t \to -\infty} Z'(t) = \infty,$$

obviously a contradiction to Lemma 89.

(ii) We treat the case $\alpha > 0$: Like above we obtain

$$\lim_{t \to -\infty} Z'(t) = -\infty.$$

Thus there exists $M > 0$ such that $Z'(t) < -M$ for every $t < t_0$. Due to $Z(0) = \nu_M - \delta$ we obtain

$$\lim_{t \to -\infty} Z(t) = \infty,$$

a contradiction to (4.108).

It remains to show uniqueness:
Let Z_1, Z_2 be two monotonically decreasing solutions of (4.107), in particular $X := Z_1 - Z_2$ solves by the mean–value theorem

$$X'' - \sigma'(\xi)X = 0,$$
$$X(0) = 0, \quad \lim_{t \to -\infty} ||X''(t)|| = 0, \tag{4.109}$$

where $\xi \in [Z_1, Z_2]$. Note, that $\xi \in (\nu_M - \delta, \infty]$ for every $t \in (-\infty, 0)$. By $\lim_{t \to -\infty} X(t) = 0$ we obtain for an arbitrary $\varrho > 0$ the existence of t_1 with

$$||X(t)|| < \varrho$$

for every $t < t_1$. The weak maximum principle yields

$$||X(t)|| < \varrho \quad \text{for every } t \in (t_1, 0).$$

Because $\varrho > 0$ was chosen arbitrary we obtain $Z_1(t) = Z_2(t)$ for every $t \in (-\infty, 0)$. Standard ODE theory applied to (4.109) yields the result. $\qquad \square$

4.6.1 The linearized Problem

Let z_1^n, z_2^n be two global minimizers of (P_{ε_n}) and let z_n be a solution of (4.17) such that $z_n \in [z_1^n, z_2^n]$. We study the eigenvalue problem

$$\varepsilon_n v_n'' - \sigma'(z_n)v_n = \mu_n v_n,$$
$$v_n'(0) = v_n'(1) = 0. \tag{4.110}$$

Let μ_n be the principal eigenvalue and let v_n be the corresponding eigenfunction. Without loss of generality we assume $v_n > 0$ and $||v_n||_\infty = 1$. The goal is now to study the asymptotic behavior of μ_n for $n \to \infty$. Like in the previous section we define

$$V_n(t) := v_n(a_n + \sqrt{\varepsilon_n}t) \tag{4.111}$$

for $t \in [-\frac{a_n}{\sqrt{\varepsilon_n}}, \frac{1-a_n}{\sqrt{\varepsilon_n}}]$ and V_n satisfies

$$V_n'' - \sigma'(Z_n)V_n = \mu_n V_n,$$
$$V_n'\left(-\frac{a_n}{\sqrt{\varepsilon_n}}\right) = V_n'\left(\frac{1 - a_n}{\sqrt{\varepsilon_n}}\right) = 0. \tag{4.112}$$

The following Lemma is crucial for our forthcoming analysis and the result is obtained similar to the corresponding result in [6] and [4]. Nevertheless we cannot give a precise asymptotic statement for the principle eigenvalue as $\varepsilon \searrow 0$ due to some technical difficulties caused by the poor knowledge about z_n'. This is an essential contrast to [6] and [4].

Lemma 91. *Let z_1^ε and z_2^ε be global minimizers of (P_ε). Then the following is valid: There exists $\varepsilon_0 > 0$, such that for every $\varepsilon < \varepsilon_0$ and for any solution $z_\varepsilon \in [z_1^\varepsilon, z_2^\varepsilon]$ of (4.17) we have for the principal eigenvalue of (4.110) $\mu(\varepsilon) < 0$.*

Remark 92. Note, that ε_0 can be chosen uniformly for all equilibria in $[z_1^\varepsilon, z_2^\varepsilon]$.

Proof. In contrast to [6] and [4] we have to prove by contradiction:
Let μ_n be the principal eigenvalue of (4.110), v_n the corresponding eigenfunction and let V_n be defined as in (4.111). We assume the existence of a sequence $(\varepsilon_n, z_n)_{n \in \mathbb{N}}$, $\varepsilon_n \searrow 0$, such that the function z_n is a solutions of (4.17) with $z_n \in [z_1^n, z_2^n]$ and $\mu(\varepsilon_n) \geq 0$ for every $n \in \mathbb{N}$.

Part 1: We claim:

(a) Restricting if necessary to a subsequence of $(\mu_n)_{n \in \mathbb{N}}$, we have

$$\lim_{n \to \infty} \mu_n = \mu = 0. \tag{4.113}$$

(b) The sequence $V_n \to V$ in $C^2_{\text{loc}}(\mathbb{R})$ and furthermore we obtain $V(t) = \frac{|Z'(t)|}{||Z'||_\infty}$, where Z is the unique solution of (4.107).

Let $x_n \in [0, 1]$ be the global maximum of v_n. Thus we have due to the usual necessary and sufficient conditions for maxima and the boundary conditions

$$v_n(x_n) = 1, \; v_n'(x_n) = 0 \text{ and } v_n''(x_n) \leq 0.$$

Equation (4.110) implies

$$0 \leq \mu_n v_n(x_n) = \mu_n = \varepsilon_n v_n''(x_n) - \sigma'(z_n(x_n)) v_n(x_n) \leq ||\sigma'(z_n(x_n))|| \leq C \tag{4.114}$$

for every $n \in \mathbb{N}$ and this yields (after restricting to a suitable subsequence) $\mu_n \to \mu \geq 0$. Furthermore we claim $x_n \in (a_n, b_n)$: We have

$$\sigma'(z_n(x_n)) v_n(x_n) + \mu_n v_n(x_n) = \varepsilon_n v_n''(x_n) \leq 0,$$

which implies

$$-\sigma'(z_n(x_n)) \geq \mu_n \geq 0.$$

By the shape of W we obtain $z_n(x_n) \in [\nu_1, \nu_2]$, which proves due to (4.92) $x_n \in (a_n, b_n)$. With respect to $||V_n||_\infty \leq 1$ and (4.112) we obtain by (4.114)

$$||V_n''||_\infty \leq ||\sigma'(Z_n)||_\infty + |\mu_n| \leq K$$

for every $n \in \mathbb{N}$ and one can show like in the previous section, that $(V_n)_{n \in \mathbb{N}}$ is relative compact in $C^2([-m, m])$ for an arbitrary $m \in \mathbb{N}$. Thus there exists V such that

$$V_n \to V \text{ in } C^2_{\text{loc}}(\mathbb{R}) \tag{4.115}$$

and we have to exclude $V \equiv 0$:
Let x_n be the global maximum of v_n. Because of $x_n \in (a_n, b_n)$ there exists $\tau_n \in [0, \frac{b_n - a_n}{\sqrt{\varepsilon_n}}]$

with $V_n(\tau_n) = 1$ for every $n \in \mathbb{N}$. By (4.93) we have $\tau_n \to \tau$ and because of (4.115) we obtain

$$V_n(\tau_n) \to V(\tau) = 1.$$

Therefore V is non–trivial and satisfies the equation

$$V''(t) - \sigma'(Z(t))V(t) = \mu V(t) \tag{4.116}$$

for every $t \in \mathbb{R}$. Because of $V_n > 0$ we have $V > 0$. (Note, that the assumption of existence of $t_0 \in \mathbb{R}$ with $V(t_0) = V'(t_0) = 0$ yields $V \equiv 0$). Thus V is the (unique) eigenfunction of the principle eigenvalue of the operator

$$\frac{d^2}{dt^2} - \sigma'(Z)I : H^2(\mathbb{R}) \mapsto L^2(\mathbb{R}).$$

Differentiating (4.99) yields

$$(Z')'' - \sigma'(Z)Z' = 0$$

and due to uniqueness of the eigenfunction of the principle eigenvalue (4.116) implies $\mu = 0$ and for every $t \in \mathbb{R}$ we have

$$V(t) = \frac{|Z'(t)|}{||Z'||_\infty} = \frac{-Z'(t)}{||Z'||_\infty} \tag{4.117}$$

by virtue of Lemma 89.

Step 2: Some estimates for v_n:
For every $x \in (0, a_n) \cup (b_n, 1)$ we have $\sigma'(z_n(x)) \geq 2c > 0$ and by (4.113) we deduce

$$0 = \varepsilon_n v_n'' + (-\sigma'(z_n) - \mu_n)v_n \leq \varepsilon_n v_n'' - cv_n,$$

for n sufficiently large. Thus we obtain

$$\varepsilon_n v_n'' - cv_n \geq 0,$$
$$v_n'(0) = 0 \quad \text{and} \quad v_n(a_n) = \kappa_n > 0. \tag{4.118}$$

Furthermore we define the boundary value problem

$$\varepsilon_n u'' - cu = 0,$$
$$u'(0) = 0 \quad \text{and} \quad u(a_n) = \kappa_n > 0. \tag{4.119}$$

We claim $u(x) \geq v_n(x) \geq 0$ for all $n \in \mathbb{N}$ and for every $x \in [0, a_n]$:
To prove this assertion we omit n for our convenience. Define $w := v - u$, let

$$x_0 := \min \{x \in [0, a_n] \mid w(x) = 0\}$$

and we assume $x_0 < a_n$. By (4.118) and (4.119) we have

$$\varepsilon w'' - cw \geq 0,$$
$$w(x_0) = w(a_n) = 0. \tag{4.120}$$

The weak and the strong maximum principle yields $w_{|(x_0,a_n)} < 0$ and $w'(x_0) < 0$. Again by the maximum principle we have $w_{|[0,x_0)} > 0$ and hence, by (4.120) and $w'(0) = 0$, we conclude

$$w'_{|(0,x_0)} > 0$$

contradicting the maximum principle. Hence we get $x_0 = a_n$, which proves the assertion.

One can easily calculate, that

$$u_n(x) := A \left(e^{-\frac{c}{\sqrt{\varepsilon_n}}x} + e^{\frac{c}{\sqrt{\varepsilon_n}}x} \right)$$

is a solution of (4.119) with a certain constant A corresponding with the boundary conditions. Therefore we obtain

$$u_n(a_n) = \kappa_n = A \left(e^{-\frac{c}{\sqrt{\varepsilon_n}}a} + e^{\frac{c}{\sqrt{\varepsilon_n}}a} \right) \geq A e^{\frac{c}{\sqrt{\varepsilon_n}}a}.$$

With x_c defined by (4.94) and the inequality $\|a_n - x_c\| \leq C\sqrt{\varepsilon_n}$ we obtain the following estimate with a constant K independent of ε_n:

$$A \leq e^{-\frac{c}{\sqrt{\varepsilon_n}}a_n} \leq e^{-\frac{c}{\sqrt{\varepsilon_n}}(a_n-x_c)} e^{-\frac{c}{\sqrt{\varepsilon_n}}x_c} \leq K e^{-\frac{c}{\sqrt{\varepsilon_n}}x_c}. \tag{4.121}$$

Hence we deduce

$$u_n(x) \leq K e^{-\frac{c}{\sqrt{\varepsilon_n}}x_c} \left(e^{-\frac{c}{\sqrt{\varepsilon_n}}x} + e^{\frac{c}{\sqrt{\varepsilon_n}}x} \right) = K \left(e^{-\frac{c}{\sqrt{\varepsilon_n}}(x+x_c)} + e^{\frac{c}{\sqrt{\varepsilon_n}}(x-x_c)} \right) =$$
$$= K \left(e^{\frac{c}{\sqrt{\varepsilon_n}}(x-x_c)} e^{-\frac{2c}{\sqrt{\varepsilon_n}}x} + e^{\frac{c}{\sqrt{\varepsilon_n}}(x-x_c)} \right) \leq K e^{\frac{c}{\sqrt{\varepsilon_n}}(x-x_c)}. \tag{4.122}$$

This implies

$$0 \leq v_n(x) \leq u_n(x) \leq K e^{\frac{c}{\sqrt{\varepsilon_n}}(x-x_c)} \tag{4.123}$$

for every $x \in (0, a_n)$. We remark that one can similarly prove

$$0 \leq v_n(x) \leq u_n(x) \leq K e^{\frac{c}{\sqrt{\varepsilon_n}}(x_c-x)} \tag{4.124}$$

for every $x \in (b_n, 1)$. Furthermore we get by the fundamental theorem in calculus and (4.110)

$$0 < \varepsilon_n v'_n(x) = \varepsilon_n \int_0^x v''_n(s)ds = \int_0^x \sigma'(z_n)v_n + \mu_n v_n dx \leq$$
$$\leq Cx \|v_n\|_{C^0(0,x)} \leq C \left\| e^{\frac{c}{\sqrt{\varepsilon_n}}(x-x_c)} \right\|_{C^0(0,x)} = C e^{\frac{c}{\sqrt{\varepsilon_n}}(x-x_c)} \tag{4.125}$$

for every $x \in (0, a_n)$. An analogues estimate holds for $x \in (b_n, 1)$.

Step 3:   Calculating the limit of $\frac{\mu_n}{\sqrt{\varepsilon_n}}$:

The following two equations are valid (in the first one we denote $w_n := z'_n$):

$$\varepsilon_n w''_n - \sigma'(z_n)w_n = b,$$
$$\varepsilon_n v''_n - \sigma'(z_n)v_n = \mu_n v_n.$$

Multiplying the first one by v_n, the second one by w_n and subtracting both, we obtain

$$\varepsilon_n(w_n'' v_n - v_n'' w_n) = bv_n - \mu_n v_n w_n. \tag{4.126}$$

Taking $\delta > 0$ fixed but arbitrary, integration of (4.126) over the interval $(a_n - \delta, b_n + \delta) =: D_\delta^n$ yields by integration by parts and division through $\sqrt{\varepsilon_n}$

$$\sqrt{\varepsilon_n}\,(v_n' w_n - w_n' v_n)\,|_{a_n-\delta}^{b_n+\delta} = \frac{1}{\sqrt{\varepsilon_n}} \int_{D_\delta^n} bv_n - \mu_n v_n w_n \, dx. \tag{4.127}$$

We intend to take the limit in (4.127) for $n \to \infty$. The estimates of Step 2 will be very useful for this procedure:

(i) Due to (4.93), (4.125) and the fact that $(\sqrt{\varepsilon_n} z_n')_{n \in \mathbb{N}}$ is uniformly bounded, we obtain

$$|\sqrt{\varepsilon_n} z_n' v_n'|_{|a_n-\delta} = |\sqrt{\varepsilon_n} z_n'|\,|v_n'|_{|a_n-\delta} \le K \frac{1}{\varepsilon_n} e^{\frac{c}{\sqrt{\varepsilon_n}}(a_n - x_c - \delta)} =$$

$$= K\frac{1}{\varepsilon_n} e^{\frac{c}{\sqrt{\varepsilon_n}}(a_n - x_c)} e^{-\frac{c}{\sqrt{\varepsilon_n}}\delta} \le K\frac{1}{\varepsilon_n} e^{-\frac{c}{\sqrt{\varepsilon_n}}\delta} \to 0$$

for $n \to \infty$. Moreover we can prove by (4.93), (4.123) and the fact that $(\varepsilon_n z_n'')_{n \in \mathbb{N}}$ is uniformly bounded

$$|\sqrt{\varepsilon_n} z_n'' v_n|_{|a_n-\delta} \le \left(|\varepsilon_n z_n''| \frac{1}{\sqrt{\varepsilon_n}} |v_n| \right)_{|a_n-\delta} \le$$

$$\le K\frac{1}{\sqrt{\varepsilon_n}} e^{\frac{c}{\sqrt{\varepsilon_n}}(a_n - x_c)} e^{-\frac{c}{\sqrt{\varepsilon_n}}\delta} \to 0$$

for $n \to \infty$.
The remaining terms of the left side of (4.127) are estimated in the same manner and all in all we obtain

$$\sqrt{\varepsilon_n}\,(v_n' w_n - w_n' v_n)\,|_{a_n-\delta}^{b_n+\delta} \to 0 \tag{4.128}$$

for $n \to \infty$.

(ii) We claim

$$\lim_{n\to\infty} \frac{1}{\sqrt{\varepsilon_n}} \int_{D_\delta^n} bv_n \, dx = \int_{-\infty}^{\infty} b(x_1) V(t) \, dt = \int_{-\infty}^{\infty} b(x_1) \frac{|Z'(t)|}{\|Z'\|_\infty} \, dt, \tag{4.129}$$

where x_1 is the unique solution of $\int_x^1 b(\tau)\, d\tau + t_r = \sigma(\nu_m)$. Note, that the last identity in (4.129) is true by (4.117). By the substitution $x = a_n + \sqrt{\varepsilon_n} t$ we obtain

$$\frac{1}{\sqrt{\varepsilon_n}} \int_{D_\delta^n} bv_n \, dx = \int_{-\frac{\delta}{\sqrt{\varepsilon_n}}}^{\frac{b_n - a_n + \delta}{\sqrt{\varepsilon_n}}} b(a_n + \sqrt{\varepsilon_n} t) V_n(t) \, dt =$$

$$= \left(\int_{-\frac{\delta}{\sqrt{\varepsilon_n}}}^{0} + \int_{0}^{\frac{b_n - a_n}{\sqrt{\varepsilon_n}}} + \int_{\frac{b_n - a_n}{\sqrt{\varepsilon_n}}}^{\frac{b_n - a_n + \delta}{\sqrt{\varepsilon_n}}} \right) b(a_n + \sqrt{\varepsilon_n} t) V_n(t) \, dt =: \tag{4.130}$$

$$= J_1 + J_2 + J_3.$$

First we treat J_1: The estimate (4.123) yields for $t \in (-\frac{\delta}{\sqrt{\varepsilon_n}}, 0)$

$$\|V_n(t)\| = \|v_n(a_n + \sqrt{\varepsilon_n}t)\| \leq Ke^{\frac{c}{\sqrt{\varepsilon_n}}(a_n + \sqrt{\varepsilon_n}t - x_c)} =$$
$$= Ke^{\frac{c}{\sqrt{\varepsilon_n}}(a_n - x_c)}e^{ct} \leq Ke^{ct}.$$

Thus we get

$$\|b(a_n + \sqrt{\varepsilon_n}t)V_n(t)\| \leq Ke^{ct}. \tag{4.131}$$

Hence we have calculated an integrable majorante for $t \in (-\infty, 0)$ and with Lebesgue's theorem we conclude

$$\lim_{n\to\infty} \int_{-\frac{\delta}{\sqrt{\varepsilon_n}}}^{0} b(a_n + \sqrt{\varepsilon_n}t)V_n(t)dt = \int_{-\infty}^{0} b(x_1)V(t)dt.$$

By the same reasons we obtain

$$\lim_{n\to\infty} \int_{\frac{b_n - a_n}{\sqrt{\varepsilon_n}}}^{\frac{b_n - a_n + \delta}{\sqrt{\varepsilon_n}}} b(a_n + \sqrt{\varepsilon_n}t)V_n(t)\,dt = \int_{\alpha}^{\infty} b(x_1)V(t)\,dt,$$

where $\frac{b_n - a_n}{\sqrt{\varepsilon_n}} \to \alpha \geq 0$ (up to a subsequence) by (4.93). It remains to consider the interval $(0, \frac{b_n - a_n}{\sqrt{\varepsilon_n}})$: Again by (4.93) we can pass trivially to the limit by Lebesgue's convergence theorem, because the sequence $\left(b(a_n + \sqrt{\varepsilon_n}t)V_n(t)\right)_{n\in\mathbb{N}}$ is uniformly bounded on bounded intervals. Thus (4.129) is established.

(iii) Furthermore we obtain

$$\lim_{n\to\infty} \int_{D_\delta^n} v_n z_n'\,dx = \int_{-\infty}^{\infty} V(t)Z'(t)\,dt < 0 \tag{4.132}$$

by the same arguments as above.

Step 4: Equation (4.127) and Step 3 yields the following asymptotic behavior:

$$\lim_{n\to\infty} \frac{\mu_n}{\sqrt{\varepsilon_n}} = \lim_{n\to\infty} \frac{1}{\int_{D_\delta^n} v_n w_n dx} \left(\frac{1}{\sqrt{\varepsilon_n}} \int_{D_\delta^n} bv_n dx - \sqrt{\varepsilon_n} \left(v_n' w_n - w_n' v_n\right)\big|_{a_n-\delta}^{b_n+\delta} \right) =$$
$$= \frac{\int_{-\infty}^{\infty} b(x_1)V(t)dt}{\int_{-\infty}^{\infty} V(t)Z'(t)dt} =: -K < 0, \tag{4.133}$$

because the denominator is strictly smaller than zero (see Lemma 89 and V is nontrivial by Step 1 of the proof of Lemma 91) and the numerator is obviously strictly bigger than zero. But (4.133) contradicts our assumption, that $\mu_n \geq 0$ for every $n \in \mathbb{N}$. $\square$

Remark 93. One can actually prove a stronger result:
Let $z_1^\varepsilon, z_2^\varepsilon$ be global minimizers of (P_ε) and furthermore we assume that (4.93) holds. Then there exists a constant $\beta > 0$ and $\varepsilon_0 > 0$, such that for every $\varepsilon < \varepsilon_0$ and for every solution z_ε of (4.17) with $z_\varepsilon \in [z_1^\varepsilon, z_2^\varepsilon]$ we have for the principal eigenvalue of (4.110) $\mu(\varepsilon) \leq -\beta\sqrt{\varepsilon}$.

4.6.2 Necessary condition for uniqueness

In this section we state a necessary condition for uniqueness of the global minimizer of (P_ε) for ε sufficiently small by considering the corresponding dynamical system to (4.17). In particular, we study the ω–limit set of the equation. Note, that the non–trivial part of the proof was done in the previous section. This part is more or less just a copy of the proof of Theorem 4 in [6]. We remark, that the same proof can be found in part 6 of [4].

As in the previous section let z_1^n, z_2^n be global minimizers of (P_{ε_n}) and we assume without loss of generality $z_1^n < z_2^n$. We denote by z_n an arbitrary solution of (4.17) such that $z_n \in [z_1^n, z_2^n]$. Furthermore we consider the reaction–diffusion equation

$$
\begin{aligned}
z_t &= \varepsilon_n z'' - \sigma(z) + \int_x^1 b\,d\tau + t_r, \\
z'(t,0) &= z'(t,1) = 0 \qquad \text{for } t > 0, \\
z(0,x) &= z_0(x) \qquad \text{for } x \in [0,1],
\end{aligned}
\tag{4.134}
$$

and we define

$$
X := \left\{ z_0 \in H^1(0,1) \mid z_1^n \leq z_0 \leq z_2^n \right\}.
\tag{4.135}
$$

Lemma 94. *The set X is positively invariant for* (4.134). *In particular,* (4.134) *defines a dynamical system on the space X.*

Proof. The proof of this statement is a standard application of the maximum principles valid for parabolic equations, which can be found for example in [36]. Nevertheless we give the proof for the reader's convenience.

For the rest of the proof we omit n. Let $z_0 \in X$ and let S be the semiflow generated by the operator $\varepsilon \frac{d^2}{dx^2}$, then we obtain by standard theory (see for example [39]) a unique solution $z(t) := S(t)z_0$ of (4.134) for $t \in [0,\infty)$. Moreover we have $z(t) \in H^2(0,1)$ for each $t \in (0,\infty)$. In particular, the solution is continuous differentiable with respect to the spatial variable.

Let us assume, that X is not positively invariant. Hence there exists $T < \infty$ and $\bar{x} \in [0,1]$, such that (without loss of generality) $z(T,\bar{x}) - z_2(\bar{x}) = M > 0$. Due to the fact that z_2 is a stationary solution of (4.134), we obtain by the mean–value theorem

$$
\begin{aligned}
(z - z_2)_t &= \varepsilon(z - z_2)'' - \sigma(z) + \sigma(z_2) = \\
&= \varepsilon(z - z_2)'' - \sigma'(\xi)(z - z_2).
\end{aligned}
$$

We define $v := z - z_2$ and $\sigma'(\xi) := c$. The above equation yields

$$
\begin{aligned}
\varepsilon v'' - cv - v_t &= 0, \\
v'(t,0) &= v'(t,1) = 0 \quad \text{for } t > 0, \\
v(0,x) &= z_0(x) - z_2(x) \leq 0 \quad \text{for } x \in [0,1].
\end{aligned}
\tag{4.136}
$$

Furthermore we define $\tilde{v}(t,x) := e^{-\lambda t} v(t,x)$ and we get $\tilde{v}_t = -\lambda \tilde{v} + e^{-\lambda t} v_t$. Multiplying (4.136) by $e^{-\lambda t}$ yields

$$
\begin{aligned}
\varepsilon \tilde{v}'' - (c + \lambda)\tilde{v} - \tilde{v}_t &= 0, \\
\tilde{v}'(t,0) &= \tilde{v}'(t,1) = 0 \quad \text{for } t > 0, \\
\tilde{v}(0,x) &= z_0(x) - z_2(x) \leq 0 \quad \text{for } x \in [0,1].
\end{aligned}
\tag{4.137}
$$

Because of $c(x) \geq C$ for every $x \in [0,1]$, where C is some constant, we choose $\lambda > 0$ such that $c(x) + \lambda > 0$ for all $x \in [0,1]$.

Due to our assumption there exists a positive maximum in $[0,T] \times [0,1]$ of v respectively $\tilde{v}$, say at $(\bar{t}, \bar{x})$. We distinguish between two cases:

(i) Let $\bar{x} \in (0,1)$:

In this case, [36], Theorem 5 in Chapter 3 yields $\tilde{v}(t,x) \equiv M > 0$ for every $(t,x) \in (0,T) \times (0,1)$. With respect to the definition of $\tilde{v}$ we obtain $v(t,x) = Me^{\lambda t}$ for every $(t,x) \in (0,T) \times (0,1)$ and continuity of v at $t = 0$ implies

$$v(0,x) = z_0(x) - z_2(x) \equiv M > 0$$

for every $x \in [0,1]$ contradicting $z_0 \in X$.

(ii) Let $\bar{x} \in \{0,1\}$ and we assume without loss of generality $\bar{x} = 1$: Again by [36], Theorem 6 in Chapter 3 we get $\tilde{v}'(\bar{t},1) > 0$, which yields $v'(\bar{t},1) > 0$ and therefore we have

$$z_0'(\bar{t},1) - z_2'(1) > 0$$

in contrast to the boundary–condition.

$\square$

Let

$$V(z) := \int_0^1 \frac{\varepsilon}{2} z'(x)^2 - F(x, z(x)) dx$$

and F is defined by

$$F(x,z) := \int_0^z \left(-\sigma(s) + \int_x^1 b\, d\tau + t_r \right) ds.$$

By the following simple calculation we obtain, that V is a Lyapunov–function:

$$\overset{\bullet}{V} = DV(z)\overset{\bullet}{z} = \int_0^1 \varepsilon z_x z_{xt} + \left(\sigma(z) - \int_x^1 b\, d\tau - t_r \right) z_t\, dx =$$

$$= \int_0^1 -\varepsilon z_{xx} z_t + \left(\sigma(z) - \int_x^1 b\, d\tau - t_r \right) z_t\, dx = \int_0^1 -z_t^2 dx \leq 0.$$

We consider the ω–limit set of z_0:

$$\omega(z_0) := \{ v \mid \text{ there exists } (t_n)_{n \in \mathbb{N}} \text{ with } t_n \to \infty \text{ and } S(t_n)z_0 \to v \text{ in } H^1(0,1) \}.$$

Since V is bounded from below in X and $V(S(t)z_0)$ is monotonically decreasing in t,

$$e(z_0) := \lim_{t \to \infty} V(S(t)z_0) \text{ exists.}$$

Moreover S is a compact semiflow (because it is generated by $\varepsilon \frac{d^2}{dx^2}$), which causes the set $\omega(z_0)$ to be compact and positive invariant in X (see for example [39], Chapter 2, Lemma 23.8). It is a well–known fact, that the ω–limit set consists of stationary solutions of (4.134) and we briefly give the reasons for that:

Let $\varphi \in \omega(z_0)$ and because of the positive invariance of the limit set we obtain $S(t)\varphi \in \omega(z_0)$ for every $t \geq 0$. Furthermore we have

$$\omega(z_0) \subseteq \{\psi \in X \mid V(\psi) = e(z_0)\}.$$

Hence by positive invariance of $\omega(z_0)$ we get $V(\varphi) \equiv V(S(t)\varphi) \equiv e(z_0)$ for every $t \geq 0$. We conclude

$$0 = \frac{\partial}{\partial t} V(S(t)\varphi) = -\int_0^1 \left(\frac{\partial}{\partial t} \|S(t)\varphi\|^2 \right) dx$$

and thus $\frac{\partial}{\partial t} S(t)\varphi = 0$.

Now we state the main theorem of this section:

Theorem 95. *We assume that for arbitrary $\varepsilon_0 > 0$ exists $\varepsilon < \varepsilon_0$ such that (P_ε) possesses at least two global minimizers $z_1^\varepsilon, z_2^\varepsilon$ and we assume without loss of generality $z_1^\varepsilon < z_2^\varepsilon$. Furthermore there exists points $a_j^\varepsilon, b_j^\varepsilon$, $j = 1, 2$, defined like in (4.92). Then we obtain for arbitrary $M > 0$ a certain ε_M with*

$$\|a_1^{\varepsilon_M} - b_2^{\varepsilon_M}\| > M \sqrt{\varepsilon_M}. \tag{4.138}$$

Proof. We prove by contradiction: We assume the existence of a sequence $(\varepsilon_n)_{n \in \mathbb{N}}$ with $\varepsilon_n \searrow 0$ such that z_1^n and z_2^n are two global minimizers of (P_{ε_n}). Furthermore we have a certain $M > 0$, such that

$$\|a_1^{\varepsilon_n} - b_2^{\varepsilon_n}\| \leq M \sqrt{\varepsilon_n}$$

holds for every $n \in \mathbb{N}$. In particular, Lemma 91 is valid for an arbitrary stationary solution $z_n \in [z_1^n, z_2^n]$ of (4.134). Let μ_n be the corresponding principle eigenvalue of (4.110) and by Lemma 91 we obtain

$$\mu_n < 0 \tag{4.139}$$

for n sufficiently large.

Let Q be the set of all stationary solutions in X. Because of

$$\dot{V}(z) = \int_0^1 -z_t^2 \, dx \quad \text{and} \quad \int_0^t \dot{V}(z) dt = V(z(t)) - V(z(0)) \leq 0$$

we obtain

$$Q = \{z_0 \in X \mid V(S(t)z_0) = V(z_0) \quad \text{for every } t \geq 0\}.$$

This yields Q is positive invariant and closed in X.

Let $z_0 \in Q$ and (4.17) implies

$$\|z_0\|_{H^1}^2 \leq C + (z_0', z_0')_{L^2} = C - (z_0'', z_0)_{L^2} \leq \frac{C}{\varepsilon}.$$

Thus Q is bounded and because $S(t)$ is a compact operator we obtain $S(t)Q = Q$ is compact in $H^1(0,1)$.

Due to (4.139) we know, that the principal eigenvalue of every stationary solution in X is strictly smaller than 0. This yields by a well–known result, which can be found for example in [21], asymptotic stability of every stationary solution in X. Thus the number of elements in Q is finite.

One can prove this assertion as follows: We assume, we have infinitely many stationary solutions $z_1, z_2, \ldots$ of (4.134) in X and all of them are asymptotic stable. Hence we obtain for every z_j an open neighborhood $N_j \subseteq H^1(0,1)$ with $z_j \in N_j$ and

$$\lim_{t \to \infty} ||S(t)N_j - z_j||_{H^1(0,1)} = 0.$$

Obviously we have $Q = \bigcup_j \{z_j\} \subseteq \bigcup_j N_j$ and due to the fact that Q is compact there exists $l = 1, \ldots, m$ such that

$$Q \subseteq \bigcup_{l=1}^{m} N_{j(l)}.$$

Let $z_r \in Q$ with $r \neq j(l)$ for $l = 1, \ldots, m$ and we assert $z_r \notin \bigcup_{l=1}^{m} N_{j(l)}$. If we assume the existence of N_k with $k \in \{j(1), \ldots, j(m)\}$ and $z_r \in N_k$ we get due to asymptotic stability

$$||z_r - z_k||_{H^1(0,1)} = \lim_{t \to \infty} ||S(t)z_r - z_k||_{H^1(0,1)} = 0,$$

which is obviously a contradiction to $r \neq j(l)$ for $l = 1, \ldots, m$.

Let $Q := \{z_1, \ldots, z_m\}$ and we define the domain of attraction by

$$\mathcal{A}(z_j) := \left\{ z_0 \in X \mid \lim_{t \to \infty} S(t)z_0 = z_j, \text{ limit in } H^1 \text{ sense} \right\}.$$

Due to the fact, that the semiflow is compact and $\{z_0\} \in X$ is trivially connected we obtain by [39], Lemma 23.6, that $\omega(z_0)$ is connected. Moreover we know, that the ω–limit set consists only of stationary solutions and this yields $\omega(z_0) = \{z_j\}$ for a certain $j \in 1, \ldots, m$.

We claim the following:

$$X = \bigcup_{j=1}^{m} \{\mathcal{A}(z_j) \mid z_j \text{ is stationary solution in } X\} \qquad (4.140)$$

and obviously "$\supseteq$" is trivial.

Thus it remains to prove "$\subseteq$": Let $z_0 \in X$ and by previous observations we know, that $\omega(z_0) = \{z_j\}$ for some stationary solution $z_j \in X$. Therefore we deduce the existence of a sequence $(t_n)_{n \in \mathbb{N}}$ such that

$$\lim_{n \to \infty} S(t_n)z_0 = z_j.$$

Let us assume for a moment the existence of another sequence $(\tilde{t}_n)_{n \in \mathbb{N}}$ with $\lim_{n \to \infty} S(\tilde{t}_n)z_0 = z_k \neq z_j$. This immediately implies $\omega(z_0) \subseteq \{z_j, z_k\}$, contradicting the connection of the ω limit set and moreover this yields the assertion.

Due to the fact, that z_j is asymptotically stable for $j = 1, \ldots, m$, there exists a neighborhood N_j with $z_j \in N_j$ such that

$$\lim_{t \to \infty} ||S(t)N_j - z_j||_{H^1(0,1)} = 0.$$

In particular, every bounded set $B \subseteq N_j$ will be attracted by z_j and by virtue of [39], Lemma 23.2, we know that $\mathcal{A}(z_j)$ is an open set in $H^1(0,1)$ and therefore X is open by (4.140). Of course we have

$$\mathcal{A}(z_j) \cap \mathcal{A}(z_k) = \emptyset \text{ for } j \neq k \quad \text{and} \quad \mathcal{A}(z_j) \cap X \neq \emptyset \text{ for every } j = 1, \ldots, m.$$

This implies X is not connected. But due to (4.135) X is convex. Contradiction. $\square$

Remark 96. Unfortunately we cannot prove uniqueness of the minimizer due to the fact we were not able to verify (4.93). In particular, if $a_2^n < b_1^n$ then (4.93) is trivially satisfied by Lemma 82 and the global minimizer of (P_{ε_n}) is unique by Theorem 95. But of course, a gap between the two transition layers can occur. I believe, that this gap can also occur in [4]. Because of that they just can prove a similar statement like Theorem 95 although they assert uniqueness of the global minimizer in their paper.

Bibliography

[1] Alexander J.C.: A primer on connectivity, in Fixed Point Theory, E. Fadell and G. Fournier, eds., Springer Verlag, New York, 455–483, 1981

[2] Alikakos N.D., Bates P.W.: On the singular limit in a phase field model of phase transitions, Ann. Inst.Henri Poincare Analyse non lineaire 5, 141–178, 1988

[3] Alikakos N.D., Shaing K.C.: On the singular limit for a class of problems modelling phase transitions, SIAM J. Math. Anal., 18, 1453–1462, 1987

[4] Alikakos N.D., Simpson H.C.: A variational approach for a class of singular perturbation problems and applications, Proc. Roy. Soc. Edinburgh Sect. A, 107, 27–42, 1987

[5] Ambrosio L., Fusco N., Pallara D.: Functions of bounded variation and free discontinuity problems, Oxford Mathematical Monographs, Oxford, Clarendon Press, 2000

[6] Angenent S.B., Mallet–Paret J., Peletier L.A.: Stable transition layers in a semilinear bondary value problem, J. Differential Equations 67, 212–242, 1987

[7] Ball, J.M.: A version of the fundamental theorem for Young measures, in: Partial differential equations and continuum models of phase transitions (M.Rascle, D.Serre und M. Slemrod, Editoren), Lecture notes in Physics 359, Springer, 207–215, 1989

[8] Carr J., Gurtin M.E., Slemrod M: One–dimensional structured phase transitions under prescribed loads, Journal of elasticity 15, 133–142, 1985

[9] Carr J., Gurtin M.E., Slemrod M: Structured phase transitions on a finite interval, Arch. Rational Mech. Anal. 86, 317–351, 1984

[10] Cellina A., Colombo G.: On a classical problem of the calculus of variations without convexity assumptions, Ann. Inst.Henri Poincare Analyse non lineaire 7, 97–106, 1990

[11] Dacorogna, B.: Direct Methods in the Calculus of Variations, Springer–Verlag, 1989

[12] DiPerna R.J.: Measure–valued solutions to conservation laws, Arch. Rational Mech. Anal. 88, 223–270, 1985

[13] Dolzmann G., Hungerbühler N., Müller S.: Non–linear elliptic systems with measure–valued right hand side, Math. Z. 226, 545–574, 1997

[14] Edwards R.E.: Functional Analysis, Holt, Rinehart and Winston, 1965

[15] Ericksen J.L.: Equilibrium of bars, J. elasticity 5, 191–201, 1975

[16] Friedman A., Phillips D.: The free boundary of a semilinear elliptic equation, Trans. Amer. Math. Soc. 282, 153–182, 1984

[17] Healey T.J.: Global continuation in displacement problems of nonlinear elastostatics via the Leray–Schauder degree, Arch. Rational Mech. Anal. 152, 273–282, 2000

[18] Healey T.J., Kielhöfer H.: Global continuation via higher–gradient regularization and singular limits in forced one dimensional phase transitions, SIAM J. Math. Anal. 31, 1307–1331, 2000

[19] Healey T.J., Rosakis P.: Unbounded branches of classical injective solutions to the forced displacement problem in nonlinear elastostatics, J. Elasticity 49, 65–78, 1997

[20] Healey T.J., Simpson H.C.: Global continuation in nonlinear elasticity, Arch. Rational Mech. Anal. 143, 1–28, 1998

[21] Henry D.: Geometric theory of semilinear parabolic equations, Lecture Notes in Mathematics 840, Springer–Verlag, 1981

[22] Hungerbühler N.: Quasilinear elliptic systems in divergence form with weak monotonicity, New York J. Math. (electronic only), 5(4), 83–90, 1999

[23] James R.D.: Co–existent phases in the one–dimensional static theory of elastic bars, Arch. Rational Mech. Anal. 72, 99–140,1979/80

[24] Kielhöfer H.: Bifurcation theory. An introduction with applications to PDEs, volume 156 of Applied Mathematical Sciences, Springer, New York, 2004

[25] Kielhöfer H.: Pattern formation of the stationary Cahn–Hilliard equation model, Proc. Roy. Soc. Edinburgh Sect. A, 127, 1219–1243, 1997

[26] Kielhöfer H.: Minimizing sequences selected via singular perturbations, and their pattern formation, Arch. Rational Mech. Anal. 155, 261–276, 2000

[27] Kielhöfer H.: Critical points of nonconvex and noncoercive functionals, Calc. Var. Partial Differ. Equ. 16, 243–272, 2003

[28] Kinderlehrer D., Pedregal P.: Weak convergence of integrands and the Young measure representation, SIAM J. Math. Anal. 23, 1–19, 1992

[29] Kohn R.V., Müller S.: Surface energy and microstructure in coherent phase transitions, Comm. Pure Appl. Math. 47, 405–435, 1994

[30] Krömer S., Lilli M.: Branches of positive weak solutions of a quasilinear elliptic equation, in preperation

[31] Málek J., Nečas J., Rokyta M., Ružička M.: Weak and measure valued solutions to evolutionary PDE, Chapman Hall, London, 1996

[32] Müller S.: Singular perturbations as a selection criterion for periodic minimizing sequences, Calc. Var. 1, 169–204, 1993

[33] Müller S.: Variations models for microstructure and phase transition, in Calculus of variations and geometric evolution problems (Cetraro, 1996), Lecture Notes in Math. vol 1713, Springer, 85–210, 1999

[34] Natanson I.P.: Theory of Functions of a Real Variable, F Ungar Publishing Co, New York, 1955

[35] Pedregal P.: Parametrized measures and variational principles, Birkhäuser, 1997

[36] Protter M.H., Weinberger H.F.: Maximum principles in differential equations, Springer–Verlag, 1967

[37] Rabinowitz P.H.: Some global results for nonlinear eigenvalue problems, J. Funct. Anal. 7, 487–513, 1971

[38] Rieger M.O.: Young–measure solutions for an elasticity equation with diffusion, in EQUADIFF 99, International Conference on Differential Equations, World Scientific, River Edge, NJ, 457–459, 2000

[39] Sell G.R., You Y.: Dynamics of evolutionary equations, Springer–Verlag, 2002

[40] Tartar L.: The compensated compactness method applied to systems of conservation laws, in Systems of nonlinear partial differential equations, Ball, J.M. (ed.) NATO ASI Series, C. Reidel Publishing Col., 1983

[41] Theil F.: Young–measure solutions for a viscoelastic damped wave equation with nonmonotone stress–strain relation, Arch. Rational Mech. Anal. 144, 47–78, 1998

[42] Vainchtein A., Healey T.J., Rosakis P., Truskinovsky L.: The role of the spinodal region in one–dimensional martensitic phase transitions, Physica D 115, 29–48, 1998

[43] Vecchi I.: A note on entropy compactification for scalar conservation laws, Nonlinear Anal., Theory Methods Appl. 15, 693–695, 1990

Index

ω–limit set, 95, 96

continuation, 10, 37–39, 41, 42

Dirac measure, 10, 11, 15, 18, 22
double–well potential, 7, 23, 25
dynamical system, 57, 95

elastic bar, 7, 14, 24, 55, 57
Euler–Lagrange equation, 7, 9, 12, 23–25,
        30, 34, 37, 38, 57, 58, 83, 88

fundamental theorem of Young measures,
        13

global bifurcation analysis, 23, 24
global continuum, 29
global implicit function theorem, 28
growth condition, 16, 29, 33

Leray–Schauder degree, 23, 28

maximum principle, 25, 37–40, 42, 45, 69,
        89, 92, 95

nonconvex variational problem, 7, 23, 57
nonlinear elasticity, 23

one–well potential, 24, 26, 36

parabolic PDE, 95

Rabinowitz alternatives, 28
Radon–measure, 12

singular perturbation, 9, 23–25, 56–58
stable solution, 10, 55, 56, 61–63, 65, 68,
        70, 80

theorem of Helly, 45, 63
transition layer, 10, 79, 84

Young measure, 11–15, 17–21
Young measure solution, 9, 12, 16, 17